THE FRACTAL SELF

The Fractal Self

SCIENCE, PHILOSOPHY, AND THE
EVOLUTION OF HUMAN COOPERATION

John L. Culliney and David Jones

 University of Hawai'i Press ◆ Honolulu

Printed in the United States of America
23   22   21   20   19   18   6   5   4   3   2   1

Library of Congress Cataloging-in-Publication Data

Names: Culliney, John L., author. | Jones, David Edward, author.
Title: The fractal self : science, philosophy, and the evolution of human
    cooperation / John L. Culliney and David Jones.
Description: Honolulu: University of Hawai'i Press, [2017] | Includes
    bibliographical references and index.
Identifiers: LCCN 2017000472 | ISBN 9780824866617 (cloth; alk. paper)
Subjects: LCSH: Cooperativeness. | Evolution. | Emergence (Philosophy) |
    Complexity (Philosophy)
Classification: LCC B818 .C85 2017 | DDC 116—dc23
LC record available at https://lccn.loc.gov/2017000472

ISBN 978-0-8248-7930-3 (pbk.)

CONTENTS

Self and Interdependence

First a note on a key term in our book's title. People sometimes misread or misinterpret the word "fractal," thinking it is a fancy way to say "fractured." But the meanings are quite the opposite. Fractal refers to profound continuity and connectedness that in mathematical formulations may propagate infinitely. In this book you will read about numerous examples of fractal organization in the systems of nature and, in human society, discover the powerful potential of the kind of person we call a fractal self. Please read on.

The story that unfolds on these pages tells how complex structure and function, and ultimately life itself, arose by virtue of the astonishing power of cooperation between *self* and *other*. Along the way, as our readers will discover, cooperation has proved a catalytic force amid some perhaps surprising conditions of universal being such as turbulence and fractal geometry. We will first explore how those simple physical states have fostered mutualism and emergence in nature, and subsequent chapters will follow this thread deep into the evolution and ethos of human nature.

Some of these phenomena and forces sound like esoteric philosophical abstractions, with vaguely mathematical and scientific overtones. We define them for our readers in accessible terms and trace their influence through evolutionary time to the present. On this journey, cooperation will loom ever larger and more clearly as a cardinal guiding principle of evolution in its physical, chemical, biological, and social manifestations. Today, science is rediscovering these concepts and integrating them with Darwinian theory. We also describe a new framework of evolutionary thought emerging from philosophy and embracing, in its formative roots, a number of ancient traditions, especially those of East Asia. Integrating this new science and rediscovered philosophy, we arrive at surprising prescriptions of how the evolutionary sense of cooperation may optimally shape our lives and open up a hopeful pathway to the future.

Both cooperation and competition were in Darwin's mind, but he and his followers greatly emphasized competition in shaping his great theory of evolution by natural selection. We have focused on new discoveries and ancient wisdom regarding a universal process in which hard edges vanish and disparate structures and functions meld together seamlessly in myriad ways across a range of scale from the atomic to the cosmic. In our book, we have gathered a wide spectrum of thought that has just now begun to pass through the powerful lens of a new scientific revolution focused on the self-organizing universe. This approach is illuminating the integrated properties and behaviors of the systems of nature, a way of knowing that philosophers and scientists call holism, which probably traces as far back as the Stone Age. Holism was adopted and developed by the ancient sages of East Asia and the Greek nature philosophers. Holistic thinking opens a window on the universe that is complementary to the perhaps later tradition of reductionism—the prevailing scientific convention that seeks to explain the whole by studying its parts. In the pages that follow, this story opens with a primer on the strangely pregnant patterns and creative powers hidden in the seamless contours of fractal geometry, strange attractors, and the edge of chaos. We then look back in time to our cosmic origins as illuminated by modern science and also as envisioned in strikingly holistic creation myths of various ancient peoples. From these beginnings we proceed on interlaced pathways, scientific and humanistic, toward new perspectives on evolution that, we believe, offer glimpses of a revolutionary synthesis of nature and human nature.

It is our view that the recent formulators of chaos and complexity theories have swept up and crystallized in computational and graphic terms a long and venerable, but chronically suppressed, way of thinking about the origin of the physical universe and its evolutionary patterns, leading to the emergence of life and human society. In this book, we traverse naturalistic philosophies and religions: pre-Socratic philosophy, Confucianism, Daoism, Buddhism, and the mythology of Hawaiian and other premodern cultures. Ultimately, we find a compelling convergence with modern holistic scientific theory—a synthesis that points not only to the evolution of mind-from-nature, but also to a realization that this emergent phenomenon has reached a seminal threshold with the evolution of human beings. In our self-organizing universe, for the first time, conscious creation has appeared. Constructive and destructive behavior in a very small part of the universe is only now backed by purpose. This changes everything.

Perhaps the most striking and frequent marker on this book's less-traveled path to an understanding of evolution and our place in it is *emergence*. This holistic state of being delights many philosophers and scientists. Its existence was first predicted in formal terms by mathematicians. Emergence is a non-Darwinian phenomenon that results in unpredictable and often powerful new properties or behaviors of systems that develop from interconnecting units such as atoms, polymers, cells, organisms, ecosystems, economies, societies, and so forth. Cooperation in nature is often a trigger for emergence. Around the turn of the twenty-first century, cooperation began to be recognized for its enormous role in shaping biotic evolution. Under the Darwinian microscope, cooperation occurs at random and is then selected for its performance in a competitive milieu. Complexity theory hints at a larger idea. Cooperation, indeed, has long functioned as a front-loader for natural selection, but it is also an expected phenomenon in the universe—a prime attractor that has sometimes catalyzed profound emergent events. Cooperation and competition may frequently oscillate, as do yin and yang. We suggest that virtually throughout evolution, cooperation has held a tiny edge over competition, an advantage probably impossible to measure accurately but one that may represent what we envision as a *cooperative constant* operating in the universe at large.

Out of the evolutionary processes leading to complexity in nature, human beings appear to have reached the potential of achieving a seminal state of being in the world—a state we call the *fractal self.** Like many other emergent phenomena, a fractal self develops and thrives in conditions that are mildly chaotic: a realm of turbulence with hidden patterns and creative potential. The principle of cooperation is closely aligned with a fractal self, that is, any person drawn to some walk of life, vocation, or avocation who begins to realize a seamless participatory ethos, as a "natural" or an "adept" with growing sensitivity, adaptation, understanding, and expertise. And such a self tends to develop a capacity to foster creative complexity that may lead to positive emergence.

This capacity, however, must be nurtured to overcome self- and society-destructive tendencies if humans are to play a catalytic role in the process of universal evolution. We currently stand at a juncture where no being has stood before. Our world of complex economic, political, and ecological processes is inexorably interwoven as never before imagined. The arrival on this scene of the fractal self points to ways to affect evolution in a positive, hopeful manner. Developing sensitivity vis-à-vis the world, such a self has the potential for shaping a future that holds a burgeoning "ecology of hope." Hence, we discover the fractal self that seeks inclusive participation in nature, as a form of saving grace for our planet and ourselves. This holistic tendency then becomes not only a philosophical and scientific issue, but also a spiritual standpoint that projects itself to a benign future.

In raising ideas at the interface of philosophy and science, our language is highly metaphorical. Metaphor has traditionally been more at home in philosophical discourse than in scientific reporting. Yet metaphor has found wide and effective use among scientists. We believe this literary device is especially useful in the early stages of grappling with concepts the logical structures and operating principles of which are partially veiled. Like electrons scattering through a molecular crystal, metaphors can help build up images that lead progressively toward understanding as we probe half-hidden contours of natural phenomena.

Borrowing freely from science, philosophy, and religion, we presume to hope that readers will encounter a newly consilient and satisfying worldview that places humanity, as fractal selves, essentially within the process of the becoming of the universe.

---

* We coined this term in an article published in 1999. See "The Fractal Self and the Organization of Nature: The Daoist Sage and Chaos Theory," in *Zygon: Journal of Religion and Science* 34, no. 4 (1999): 643–654.

ACKNOWLEDGMENTS

No book is ever written alone, especially one titled *The Fractal Self.* We are grateful to so many family members, friends, colleagues, and students over the years who have listened to us and contributed to our dialogue with each other. After years of writing, discussing, and presenting these ideas at conferences and forums, listing everyone is impossible.

As coauthor whose main contributions in the book describe scientific views of evolution and emergence in the universe from the big bang to sociobiology, John would like to thank Lane Yoder for key insights on chaos and fractal geometry. Ted LaRosa provided essential guidance in the chapter on cosmology, as did Eric Vetter in discussions of the ecology of cooperation. The work of primatologist Frans de Waal has been central to our envisioning of the evolution of the prototype of a fractal self. He graciously reviewed chapter 6, which focuses on the emergence of conscious cooperation and the advent of morality. David Mumford, Caroline Series, David Wright, James Cheetham, Margaret McFall-Ngai, Zanna Clay, and Renzo Lucioni generously gave permission to reprint images from their researches, ranging from a fractal Indra's Net to bonobos embracing to political disjunction in the US Senate. Susan Culliney's drawings enrich the narrative and enhance understanding of key topics in several chapters.

For conversations on ideas arising as the manuscript emerged and/or comments on clarity of expression in his writing, John is happy to thank Aaron Culliney, Susan Culliney, Barbara Mayer, David Weir, Ann Kircher, and Nan Sumner-Mack.

Much of the biology in this book came out of John's teaching. He is grateful to numerous students in those classes over many years, as well as faculty colleagues, for deep-searching discussions into evolution and how we come to know about it. If they read *The Fractal Self,* they may find passages that recall their contributions to ideas and metaphors that emerge in these pages.

Among other sources of inspiration, John wishes to acknowledge the writings of Patrick O'Brian and Kim Stanley Robinson, historical novelists of ecopoetic sensibilities across scale. Many of their stories bracket our time on Earth and dramatically voice the deeply humanistic, cooperative vision that is both our heritage and our hope.

For those contributing directly to the book, David thanks Thomas P. Kasulis for providing not only the use of his diagrams from his landmark book, *Intimacy or Integrity: Philosophy and Cultural Difference,* but also a pivotal measure of the inspiration for our conception of the fractal self, which is a self of ascendant intimacy. Although we extend his use of intimacy, without his initial insights this book would have been differently written. Roger T. Ames was instrumental in providing feedback for chapters 7, 8, and 9. His close reading and suggestions were crucial in making these chapters more exact and richer. Robert Olby generously offered assistance on the discussion of Francis Crick in chapter 7. Long and fruitful conversations with Martin Schönfeld provided necessary corrections and trajectories of thought. And we are particularly grateful to Joseph Overton for allowing us to use him as an example of a fractal self in the last chapter.

Some of this book was written and reworked while David enjoyed several visiting scholar opportunities at the Institute for the Advanced Studies in Humanities and Social Sciences at National Taiwan University. He is most grateful to Huang Chun-chien, the institute's dean and distinguished Confucian scholar, for his support over the years and the friendship that has ensued. Dean Huang is the living embodiment of the Confucian sage and is surely a fractal self who connects people and projects and lets the course of things amplify according to their natural zeal. Kirill O. Thompson, the associate dean of the institute, is the proverbial person behind the scenes—always present yet only occasionally visible. Kirill gently makes good things happen and has become a dear friend over the years. He is another fractal self. Alice Pate, chair of the History and Philosophy Department at Kennesaw State University, has been most supportive of this project over the years and has provided what she could in times of trimmed budgets.

Much of the mythological and philosophical work contained in the *Fractal Self* has been influenced by the work and voice of Graham Parkes. It was through Graham Parkes that David gained his philosophical tenor. The power of this influence is too extensive to be individuated in any precise way, but resonates throughout this book. Additionally, we are very grateful for the generous offering of reprinted photographs by Jason Bound and Nathan Wirth.

Some of the ideas contained in this book and small portions of our chapters have been published previously. Most recently, in 2016, the *Taiwan Journal of East Asian Studies* published "In an Age of Global Decline: The Need for a Return to a World Classical Philosophy 濟世之道：回歸世界古典哲學," and "Heaven-Earth-Self: The Fractal Divinity of Self" appeared in the *Journal of Religious Philosophy* in Taiwan. This same journal also published "Letting Life In: Religious Concord through Chinese Sources" in 2014. The seed for *The Fractal Self* was planted in two articles published in *Zygon: The Journal of Science and Religion.* "Confucian Order at the Edge of Chaos: The Science of Complexity and Ancient Wisdom" was published in 1998, and "The Fractal Self and the Organization of Nature: The Daoist Sage and Chaos Theory," where we first coined the term "fractal self," appeared in 1999. This latter article was republished in the *Asian Culture Quarterly* in 2001 and translated into Chinese in *Zhongwen Zixue Zhidao* 中文自学指导 (*Journal of the Chinese Language Department of the East China Normal University*) in 2008. We are grateful to the editors of these journals for the inclusion of these materials.

Through numerous sacrifices and financial obstacles, E. R. and Marie Jones provided the gateway opportunity that amplified into a world of unimagined openings not found for many in David's working-class neighborhood. To E. R. and Marie, David dedicates his portion of *The Fractal Self.*

Both of the authors are grateful to the editors and art department staff at the University of Hawai'i Press for their strong support and attention to details in the text and illustrations. We especially thank Nadine Little for her early advocacy of the manuscript. Debra Tang and Shelby Pykare assiduously kept the book's trajectory on the deterministic side of chaos. Pam Kelley's long experience with complex projects reassured us of emergence in the end.

Introduction

From Chaos to Intimacy, A Primer

Perhaps the most fundamental question in philosophy and cosmology asks, "Why is there something instead of nothing?"[1] In this book we approach that question, but like cosmologists and physicists who seek the earliest manifestations of the universe, when all matter and fields of energy were formative, we have to start just after the moment of initial being. And so we rephrase the question: Why is there something that has become wondrously complex—for example, as (writ large) in physics giving rise to chemistry and beyond to biology—instead of staying utterly simple?

Our book is about the process of evolution in the most inclusive sense. We describe prevailing patterns of cosmic to microcosmic change in the universe that science has just begun to understand in new ways—patterns that strikingly intersect with classic insights of philosophy—particularly in the Daoist and Buddhist traditions. At their most seminal, these patterns and processes address roles of life and human nature within nature at large. Out of contemporary science and surprisingly congruent conjectures of ancient wisdom comes an understanding of why we observe structure and order in the universe and why there has arisen a long-term trend toward intricate pattern instead of universal randomness.

The universe has a powerful tendency toward self-organization. It builds complex systems embodying many interacting units that cooperate and compete in myriad ways. In opposition to simplifying, dissipating forces such as the second law of thermodynamics, an opposing principle of interaction that emphasizes cooperation over a huge range of scale has bootstrapped emergent manifestations of matter and energy in progressive leaps of complexity. Billions of stars result in the glory of galaxies; billions of neurons produce the astonishing potential of mind and self. In our book we aim to trace the giant steps of this progression. We propose that universal evolution has been catalyzed at every step by a new principle, the *cooperative constant,* and we will speculate on where the trail may yet lead.

Foundations of Being: Chaos and Complexity

Seeking a point of departure from which to follow universal evolution, scientists a generation ago began to focus on a dynamic state of nature that leads out

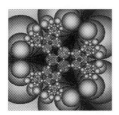

Figure I.1 Indra's Net (or Web) is one of the most striking depictions of interdependence and interpenetration in Buddhism. Shown here distributed fractally, Indra's Net stretches endlessly in all directions. Each node in the net is a jewel, and each of the infinitely many of them reflects the light of every other. Each then is interdependent and imbues the entirety that arises in this network of relations, and self-organization is at its core. *Source:* David Mumford, Caroline Series, and David Wright, "Indra's Pearls," Cambridge University Press. Used with permission. A number of artistic-fractal representations of Indra's Net can be seen at http://www.hiddendimension.com/Tutorials/Slope/Slope_Tutorial_9.html.

of *chaos,* now viewed by many scientists and mathematicians as harboring the starting condition of self-organizing complexity in nature. Chaos can set the stage—reach a sort of tipping point—for sudden pulses of change in both natural and human systems. This is the state of water molecules on the forward edge of an ocean wave an instant before the wave breaks. It is the state of stem cells in an early embryo just as it rapidly starts to take shape as a fetus; it is the state of various plant and animal species assembling a pioneering ecosystem on a newly erupted island. It is the state of a free, entrepreneurial economy, such as that which nurtured the personal computer industry at its inception.

To introduce readers who have had little or no exposure to chaos theory, we begin with a primer to provide a working familiarity with the terms and concepts that are critical to our later narrative and arguments. For those who want a grounding in these ideas at an introductory level, we recommend the eminently readable 1987 classic, *Chaos: Making a New Science,* by James Gleick.[2]

Deterministic Chaos and Attractors

"Chaos," in modern colloquial usage, actually describes situations that are not wholly chaotic. Mathematicians refer to a realm of nonlinear systemic structure and behavior that is not the same as randomness, which is the absence of pattern. They call this pattern-forming realm *deterministic chaos.* Throughout this book, unless otherwise stated, this is the kind of chaos that we will refer to—the principle behind the shaping of many complex systems. Not surprisingly, the patterns discovered in deterministic chaos are often complicated; they are typically not precisely predictable, but may be circumscribed by limits and calculated probabilities. Nonlinearity means that these sorts of chaotic systems wander as they evolve structure and function through a space of possible existence.

The pattern of structure or functionality that characterizes a system shaping itself out of deterministic chaos is an example of an *attractor.* It embodies the particular organizing principle for a given system in nature. For instance, one

Figure I.2 Lorenz Butterfly Attractor: This graphic pattern arises from equations of Edward Lorenz to describe simple modes of airflow in the atmosphere that can shift abruptly and unpredictably when small disturbances impinge on the flow. Yet the oscillating pattern is constrained; it is infinitely variable, though nonrandom, never exactly repeating its path. *Source:* https://commons.wikimedia.org/wiki/File:Lorenz_attractor2.svg.

of the simplest kinds of attractor is described by the motion of an oscillator; a planet's orbit is an example. Even though, when followed through many years, the exact path of a planet's path around the sun may vary, or wobble slightly, it typically tends to zero in from one side or another toward its natural cycle. An attractor can be graphed; it shows a three-dimensional pattern within which all real motions of the oscillator will trend.

Turbulence

Although an oscillator's average or most probable behavior can be described in the linear language of a stable state, it has been discovered that in many systems, which may vary greatly in complexity, an attractor has unstable border regions within which a system can range well beyond the bounds of equilibrium. Enough energy applied from within or without can cause a system to approach and exhibit chaotic behavior in the realm of turbulence.

Turbulence—to an engineer, physicist, oceanographer, or meteorologist—is a seemingly disorganized pattern in the flow of a fluid. Theoretically, the waveform of a modest ocean swell approaching a coast in the absence of turbulence can be reliably calculated or graphed—until the crucial state in the interval of the wave's breaking. Then minute variations in the wave's angle of approach to the shore, small irregularities in the shoaling sea-bottom, a fish passing through the base of the wave, small shifts in the wind above the sea, even the presence or absence of a surfer waiting to ride the wave, can rapidly trigger the variability in turbulence that shapes such a breaker in a unique pattern every time. The edge of a just-breaking wave, as in any fluid entering turbulence, defines a characteristic contour at the limit of the system's orderly behavior. Chaos theory shows us that this contour is a dynamic boundary of the attractor for the system. At the transition to turbulence, this boundary is known to many scientists and computer modelers as the *edge of chaos*.[3] Diving into such a contour, one sometimes finds unexpected shapes and forces at the transition to turbulence, the potential for whole new directions that may be taken by the system in question.

Sensitive Dependence on Initial Conditions

Many scientists would like to understand the behavior of more complex systems than planetary orbits and probe turbulent systems of greater interest than water waves. The global climate system provides an example. We are used to the idea of oscillation as a property of climate and do not find it strange that summer and winter in temperate and polar regions, or wet and dry periods in the tropics become parts of the space of possible climate behavior within a seasonal world attractor. The climate behaves, on average, as if the planet is a pendulum—swinging to its extremes in summer and winter.

However, the great complexity of patterns in atmospheric circulation has led scientists to an appreciation, if not a complete understanding, of one of the signatures of deterministically chaotic behavior—*sensitive dependence on initial conditions.* Indeed, the classic example involves the local weather (meaning short-term effects that emerge out of the long-term pattern of climate). In modeling weather patterns on supercomputers, meteorologists establish a set of starting points—temperature, humidity, barometric pressure, wind speed and direction, and so on, and then let the weather system evolve. In any subsequent model with simulations that allow even very slight changes in initiating conditions, wildly different virtual weather patterns emerge. This phenomenon was first named the "seagull effect" by the theoretical physicist Edward Lorenz, whose pioneering 1963 computer simulations forever dashed the hopes of the then-reigning reductionist school of weather forecasting.[4] Prior to Lorenz's experiments, it was thought that small imbalances or disturbances in weather factors would always just damp out and disappear into a larger, predictable unfolding pattern.

The Butterfly Effect

The "seagull effect" was soon replaced by a parallel expression that emphasizes sensitive dependence even more strikingly. In any complex oscillating system, if a snapshot could be taken of the system's behavior at particular phases, or sectors of a cycle, it would resemble but never exactly duplicate many such snapshots taken in subsequent cycles. Thus the boundaries of a system's attractor, along the transition to turbulence, conceal a continuous potential for quantitative and qualitative change. Revisiting the climate example, picture Cape Cod in July. Every summer's weather and landscape generally look the same, but, of course, each summer exhibits somewhat different patterns—turbulence, or roughness, on the scale of day-to-day shifts in the weather. The discovery of dependence on initial conditions in meteorology was ultimately christened the "butterfly effect" to emphasize the extreme sensitivity. In some ultra-rare instance, there might be some truth in the now hackneyed speculation that a butterfly flapping its wings in Texas could trigger a tornado three days later in Ohio.[5]

The Sandpile Effect

In a nearly forgotten set of studies, the physicists Glenn Held, Per Bak, and others confirmed theoretically and experimentally Lorenz's inference that a disturbance need not be large to have great impact.[6] "Self-organizing criticality" is the physicist's term (roughly equivalent to edge of chaos) for the development of a state of maximum sensitivity (to disturbance) toward which the structures of certain complex systems evolve. Potentially, at criticality, the addition (or loss) of a grain of sand may bring down a mountain. In experiments of extreme precision and beguiling simplicity, Held piled up sand to a threshold of instability. This is the point at which a single added grain, dropped on the apex of the pile, may trigger a partial collapse of the pile. In this situation, the initiating event looks the same every time; it appears as a unit impact, but the potential result ranges across a wide magnitude. Mostly, of course, the consequence is a tiny slide of grains that slightly improves the pile's stability, but once in a while, the result of the unit impact is a large collapse. Between the extremes is a classic curve that relates the size of the sand avalanche to the expected frequency; it follows a *power law.** It is easy to predict the frequency of any magnitude of collapse. Small sandslides are common; big ones are rare. The surprise is that they all can start with a single added grain when the pile is at the edge of its stability.

Complexity theorists are aware that the general power law governing sandpiles applies to the magnitude of change in many other systems that are subject to (often small) episodes of disturbance. The "sandpile effect" is merely a simplified version of the butterfly effect. The shock wave of each sand grain's impact, like the force imparted to the air by the insect's wings, has the potential to propagate through the pile, across its virtually seamless spectrum of scale, to trigger a major change. Of course the weather embodies much greater complexity than the sand, in the sense that there are more variables to provide negative feedback that may dissipate the butterfly effect before it goes very far in any one direction. This model then is predictive, but only in the sense that we can know the chance of small, medium, or large changes occurring.

Fractal Geometry

That an infinitesimal nudge of a complex system in a critical state could produce a large effect may not be surprising given the way the attractors of such systems are structured according to the principle of *self-similarity regardless*

---

*Frequency (probability) varies inversely with the size of the event (landslide), raised to some power. Other scientists later discovered effects that, depending on grain size and sandpile diameter, depart from this fundamental rule; some of those studies may be useful in predicting earthquakes.

of scale. Telescoped into attractors of many complex systems is a turbulent and recursive geometry, possessing profound structural order, yet bordering on the chaotic, and devolving to startlingly simple quantitative laws that mathematicians—some with famous names, such as Leibnitz and Poincaré—struggled with for more than two centuries. Their discoveries appeared to violate conventional wisdom and were often disbelieved by colleagues. Thus the strangeness of mathematical sets, and their geometrical patterns being self-similar across scale and having "non-integer dimensions," held back an appreciation of the ubiquitous presence of these patterns in nature. This was finally elucidated by Benoit Mandelbrot in the 1960s and 1970s.[7] Using the emerging power of computer graphics, Mandelbrot navigated boldly into fractal depths that had been previously unfathomable. What he found was that forms and processes in the universe seemed to repeat themselves on every scale and that nature created structure in seamless dimensionality—no longer could the real world be accurately simplified by reference to linear coordinates and whole-number dimensions. Between familiar constructs such as a two-dimensional surface and a three-dimensional volume was an infinity of intermediate form. Mandelbrot introduced the term "fractal" (the key to the fractional geometrical world). And it became a breakthrough that cracked the code of chaos. It is now clear that the infinitely iterative fractal governs the onset of turbulence on the borders of attractors throughout nature at all levels of organization.

The Fractal World

Picture a sponge. How to describe its dimensionality? If we don't think very deeply about it, and if it is a fairly dense type of sponge, we tend to call it a three-dimensional object. However, a twinge of doubt may remain as you think about the way a sponge takes up space. One might consider it to be an extremely folded (curved and crumpled) two-dimensional surface. Actually, when subjected to a mathematical analysis, a sponge turns out to have dimensionality between two and three. It takes up space in a way that is more than a surface but not yet three-dimensional. The geometry of a sponge is fractal.

Mandelbrot built on a realization of some of his unsung predecessors, notably Paul Levy, that nature was not smooth and that its patterns of roughness were not random; they repeated themselves at every scale. Mandelbrot's subsequent observations and experiments, followed by those of many other scientists, thus revealed fractal organization in all sorts of systems of varying complexity, ranging from the atmosphere, flowing water, and not-so-solid earth, such as layers of lava, to organisms to economies to electronic communications. Of course, fractal patterns are most obvious in shapes we can see—sponges, coastlines, mountain landscapes, growing plants, clouds in the sky. But chaos theory suggests such objects are shaped by forces that define dynamic attractors bordering on deterministic chaos. Always we are led back to turbulence. Most often, natural objects seem to be forms frozen out of turbulence, and the very geometry of turbulence—oscillating, branching, mixing, multiplying, folding, fragmenting—is fractal.

Chaos theory confirmed the hidden structure of systems in turbulence and demonstrated the realities of roughness and illusions of dimension in the geometry of nature. Although the theory seems to have universal applications, describes vast self-similarity in the universe that is independent of scale, and calculates fractal patterns and orderly ramifications that lie far from randomness and apparently extend to infinity, chaos theory itself seems caught in its own metaphysical attractor. It is a compelling exposition of the structure and behavior of systems as long as they do not change their basic natures. Chaos theory, like theories of equilibria, is essentially static; it traps any particular system in a defined sector of the space of possible existence. Sometimes, however, much greater potential may dwell along the boundaries of attractors than meets the chaos-adapted eye.

Complexity and Complex Systems

The science of complexity is the child of chaos theory. This approach to understanding universal evolution opened up in the 1980s and 1990s with breakthrough research by the likes of biologist Stuart Kauffman, systems scientist Norman Packard, and computer modeler Christopher Langton, among numerous others.[8] Understanding complex systems proceeds from the particular and immediate levels of system structure and performance described by chaos theory, but complexity informs the long view and seeks principles of system evolution in which ultimately the boundaries imposed by longstanding attractors may be broken or refashioned. Complexity probes a kind of metamutational physiology at the heart of a powerful engine for progressive development in universal evolution.

Biologists sometimes view ecology—the patterns of interaction between organisms and their environments— as snapshots or single frames in the long movie of biotic evolution. Perhaps there is a parallel in the relationship of chaos theory to complexity—the former constituting a collage of individual frames out of which emerges, with sometimes surprising twists, fades, and jump cuts, an epic and burgeoning saga of complexity.

The systems that are the grist of complexity theory are made up of many interacting units and subsystems of units. Examples of complex systems include national and global economies, governments, cities, corporations, and other manifestations of society; also nonhuman entities such as cells, animals, plants, and other organisms; ecosystems; and the biosphere. Inanimate nature also produces systems of this sort, for example, the oceans and atmosphere of the Earth. The interacting units themselves are as variable as atoms and molecules, living cells, people and human institutions, and the myriad species in the biosphere. Chains of related complex systems are evident in which units become hierarchical, as in biology—molecules to cells to organisms to populations to ecosystems. Unit interactions (or transactions) can be thought of as forms of communication, or information exchange, even in cases of simple chemical and physical interaction (e.g., molecular collisions and chemical reactions).

Emergence

In complexity theory, going beyond the limits of chaos, is a focus on dynamics within a complex system that can induce the system to leap off the tracks in its erstwhile space of possible behavior. It may metamorphose into a whole new state that radically warps or abandons the system's known attractors. While multiplicative or synergistic effects commonly emerge from relatively simple interactions in a variety of systems, sometimes qualitative jumps occur that are beyond imagining. Such evolutionary change that cannot be predicted from measurements of the behavior that led to the change is termed *emergent;* in its simplest context, this is where the whole becomes more than the sum of its parts.

Emergence is the definitive evolutionary outcome of complexity. Certainly there is no argument that historical accident and contingency, based on what has been set in motion or established at any given point, influence specific direction and outcome in the evolution of life (an organic butterfly effect to be sure). However, complexity theory suggests that self-organizing mechanisms intrinsic to complex systems provide initiative shaping of system structure, including living entities such as cells and organisms, and also generate impulses and guidance to system behavior that may trigger emergence. In other words, the generalized attractors for various kinds of systems and also the trend toward progressive, or directional, change may not arise entirely by chance. Stuart Kauffman has called this evolutionary process and sorting out that occurs spontaneously among interacting units "order for free."[9]

Unfortunately, misinterpretation of the emergent process is rife. To repeat: what is not indicated here is the preordination of any specific structure or organism. Turn back the clock and let evolution on earth run its course a second time.[10] Everything would be different in detail—the butterfly effect again. However, just as the average climate is shaped by prevailing attractors, complexity theory predicts the likelihood that recognizable "cells," "organisms," and "ecosystems" would assemble a biosphere, or a succession of them, that would resemble in general what we find in our living world. We might not be able to call the organisms insects or fish or mammals, yet eventually ecosystems would probably incorporate something like them. Neither would the timetable of hierarchical emergence necessarily repeat itself. For example the rerun of our world might remain exclusively microbial (or prokaryotic) for a longer or shorter period.

Back to the Edge of Chaos

According to complexity theory, a kind of progressive evolution seems to be characteristic of complex systems. They coalesce from simple precursors that may be haphazardly or weakly interactive, with little sustainable structure. A system then evolves in the direction of greater complexity that may involve approaching criticality or the edge of chaos. The edge may involve unstable structural diversity, information flow, and/or communication potential among the system's units. A high volume, diversified global economy or an ecosystem rich in species is thought to dwell close to this seminal condition of being. In this state, volatility and richness of interaction are at a sustainable peak and the system as a whole is precarious and highly sensitive to disturbance—sometimes from outside the system and sometimes from within—in the form of innovative change and self-selection among its component units and subsystems that can lead unpredictably to emergence. The edge of chaos, now a landmark in complexity theory, is the boundary of a system's attractor at the transition to turbulence, a border of possibilities for system existence. It may be of utmost importance that the potential for emergence—in the form of "creativity," or tipping points that stem from cooperative links among system units, or recombinant "seeds"—appears to reach a maximum at the edge of chaos.[11]

Beyond the edge lies deepening chaos amid disorder, where a system's oscillations wander in infinitesimal recursion, perhaps dissipating into randomness. This is a metaphorical cliff. Here a disturbance or an emergence (sometimes they occur together, or one may lead to the other) may push the system over the cliff. Much of the functional structure, sustainability, orderly flow of information, or creativity of the extant system may be lost. Mass extinctions or other catastrophes are possible; some units break their connections and disappear, while others form new links. Disturbance and emergence reset the boundaries of a system's attractors and may precipitate profound qualitative changes in the system itself.

A Cosmology of Hope

The history of the cosmos on all scales has shown us that wherever energy breaks free of some confining state (as gravitation, as sunlight, as a concentrated burst of heat, and so on) there is, at the expense of that energy, a trajectory of development toward complexity. This is what we mean by universal evolution; wherever there is potential energy to tap, nothing gets simpler, or regressive, or randomized for long. Of course, physicists and engineers point out that the universe as a whole is winding down, cooling off, dissipating its concentrated energy, and increasing in disorder all the time. The famous second law of thermodynamics again. However, raveled into the fractal structure of the cosmos itself we see the paradoxical opposite happening in myriad places where matter and energy have formed intricate constructs over a huge range of scale. For billions of years, those parts of the universe—galaxies, stars, planets, and their self-organizing systems—have gained universally in patterned structure and intricate function.

The universe is a cosmic ocean awash with innumerable waves of complexity at many frequencies. Not too much happens in the vast intergalactic deeps. It is much more interesting around the edges, skirting galaxies, stars, and planets, where roughness develops and swells of potential change begin to crest. The power of evolution is in emergence, and emergence is most likely to happen in and around turbulence, where systems are most sensitive to any small change or disturbance. However, the world is used to waiting. Arguably, numerous species of complex animals, especially vertebrates, project a kind of selfness—individuals develop flexible behaviors, and the rudiments of culture may emerge in subgroups of social mammals and birds—killer whales, elephants, and crows. And evolution of the mammals at large (when released from global dinosaurian dominance) took some sixty million years of adjustments to new ecological attractors and adaptive responses to selective pressures before a line of primates with all the right stuff—big brain, elaborate larynx, hands with thumbs, tilted pelvis, social altruism, and so on—poised itself for a giant leap. Is it likely (or inevitable) that if this hadn't happened to human ancestors, something else eventually would have found our sociobiological attractor, or a comparable one? In another sixty million years, perhaps?

Fractal Life

Fractal geometry features repetitive patterns in turbulence of the same sort that appear along the boundaries of attractors, and it is now clear that such fractal roughness, rather than the rigidity of integral dimensionality, characterizes the shaping and dynamic behavior of nearly everything in the real world. And it is in roughness that important things happen. Complex crystals and polymers grow out of roughness at the edges of basic materials. Organic chemical roughness of key molecules triggers catalysis and mutation that transform the potential of organisms. Cellular roughness led to the evolution of eukaryotes. Tectonic, atmospheric, and oceanic forces impose roughness on continents and islands, and this initiates ecological roughness that transforms species. The potential complexity in roughness that shapes the structures and behaviors of natural systems keeps getting deeper and ever more pregnant with content. Thus, as biological entities explore their way into those systems, they are continually challenged by their surrounding dynamics of nature. And then the effectiveness and fitness of their essential attributes for survival and reproduction must pass the test of natural selection.

Organisms can behave in ways that sand grains do not. Individuality of cells and organisms stems from genetics and is endlessly mutable in ways that are compounded by vicissitudes of interaction with the environment. Keystone species organize the structure and dynamics of their ecosystems. When sea otters, for example, are extirpated from a region, the kelp-forest ecosystem often collapses to a mere vestige of its normal vigor and diversity.[12] This is a long way from the sandpile effect. Yet not even sand grains are identical; occasionally, one might act as a keystone. (Perhaps we have to regress to electrons and quarks to arrive at system-building units that begin to approach uniformity.)

Human communities feature many examples in which subtraction or addition of individuals leads out of volatility to major change, chaotic rearrangements, and emergence. The arrival of people such as the Wright brothers or Martin Luther King Jr. in their particular socioeconomic-political and/or technological settings changed the world. Of course, like the ultimate sand grain, it was impossible to predict a priori that those individuals, as societal keystones, would

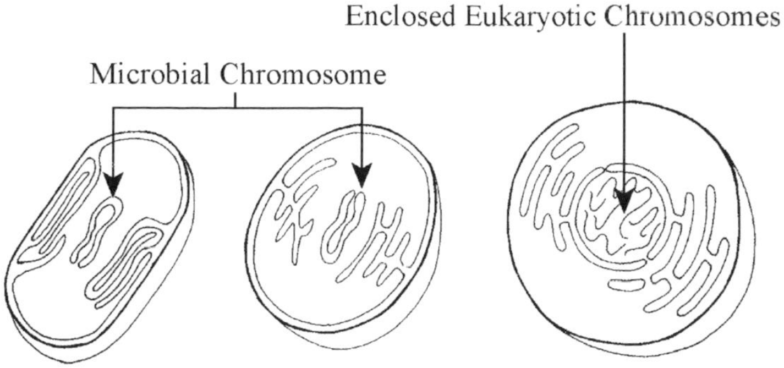

Enclosed Eukaryotic Chromosomes

Microbial Chromosome

Figure I.3 Fractal organization and roughness of cellular membranes led to the emergence of advanced (eukaryotic) life-forms. The figure shows progressive ingression of the cell's boundary (plasma membrane) to ultimately surround the central chromosomal material (DNA with accompanying proteins). In the left and middle drawings, the chromosome is a closed (prokaryote-type) loop, here drawn in an elongate horseshoe shape. The cell on the right shows the eukaryotic transformation, with DNA having fragmented into individual filamentous chromosomes of varying length within an enclosing nucleus. *Credit:* Drawing by Susan Culliney.

precipitate large episodes of change. A social application of complexity theory would suggest that various aviation pioneers in the early twentieth century and civil rights leaders of the 1960s were at work in turbulent settings, along the edges of chaos, in their respective fields. Later in this book we will arrive at a critical focus on the influences of such individuals in the world. They typically emerge from the ground up, are often cooperators, and become catalysts of significant change. While every human being has potential to change the world in some way, the examples indicated above rise to especially high profiles on the spectrum of human achievement. They are preeminent types of people that we call *fractal selves,* who immerse themselves adroitly, inserting almost seamlessly into sectors of the world that attract them and often foster emergence.

Competition and Cooperation, Intimacy and Integrity

In the long course of universal evolution, it appears evident that two opposing principles have shaped the many stages of cosmic development. Wherever encounters happen between and among the emergent forces, materials, and objects of our universe, there has perhaps never been a state of indifference. Sometimes the issue is one-sided, with no contest, but often a tension appears, as if between yin and yang, an incipient competition versus a tendency toward cooperation. These terms are usually reserved for biological and cultural levels of organization. But we have applied them across scale as useful metaphors to refer to physical, chemical, and biological systems, beginning with the cosmologists' conjecture of a "confrontation" between matter and antimatter at the dawn of cosmic existence (see chapter 1). In our discussions of competition and cooperation throughout this book, we point to evidence and thought experiments indicating that competition often leads to simplification of systems, while cooperation (or sometimes avoidance of competition) tends to make them more complex. Recently, compelling insights have confirmed the profound influence of cooperation in evolution at all levels from molecules to social systems. In part, this idea represents a renaissance and resurgence of the concept of group selection that had been glimpsed by Darwin in *The Descent of Man* (see chapter 6), but only began to accrue serious discussion in the mid-twentieth century. Then, for several decades, such studies generated controversy among evolutionary theorists. But by 2000, biologists began new explorations of cooperative evolution on many levels,[13] and never has its importance stood on firmer footing than now, braced with mathematical rigor developed by the Harvard mathematician and theoretical biologist Martin Nowak and his colleagues at various institutions around the world. Out of research and insights by Nowak's network, cooperation has rapidly emerged as a defining principle in evolution, on par with competition, adaptation, and natural selection[14] (also see chapter 6).

From cultural philosophy and psychology we know that early thinking about the self in traditions from Asia to the West considered the self to be a participatory being embedded both in nature and, intimately with others, in human society. Such a way of being and thinking is especially constituted in Confucianism, Daoism, Buddhism, and in early Greek philosophy in the West. We discuss these philosophies and traditions of the cooperative self in detail later in the book (chapters 7, 8, and 9). In our own time, gaining insights from comparative studies, the philosopher Thomas P. Kasulis, in his *Intimacy or Integrity: Philosophy and Cultural Difference,* has focused on a spectrum of cultural practice and social behavior extending between polar manifestations of integrity, the emphasis of which is individuality (with implied competition of self versus other), and intimacy, which leads to cooperative relationships as an intuitive pattern of existence and being in the world, and becoming and evolving with it.[15]

In considering evolutionary effects of competition and cooperation, an important nuance that has been realized out of research in ecology is that *intensity* is crucially important in shaping outcomes.[16] Extremely aggressive actions among competitors may be damaging to all parties, or may lead to overwhelming dominance by one or a few, reducing richness of a system. On the other side, cooperation often seems to nurture complexity in natural systems. However, intense cooperation, for example in rigid, locked-in behavior patterns of ants, drastically represses freedom of individual action or expression. In the realm of biology, mutation and sexual reproduction continually shake up the course of evolution and, with some exceptions, help keep life's trajectories away from extreme states (see chapter 5). An excess of stability or rigid regulation reduces robustness of system structure and function and minimizes chances of emergence. This again reminds us that as systems in nature self-organize, they tend to approach the edge of chaos.

Likewise, in cultural evolution, states of being that emerge, sifting expressions of integrity and intimacy, generally seek some intermediate realm along a spectrum between the extremes. Thus, as Kasulis points out, it is usually not a matter of behavior of a given system proceeding merely from one stance or the other, but more a matter of the foregrounding of intimacy in certain situations and integrity in others. In other words both are usually present to varying degrees. While these two attitudinal conditions arrive at their peak realizations in the world of social and cultural affairs, they provide compelling metaphors to represent a fundamental duality or existential oscillation from which have emerged many seminal patterns in the history of the cosmos.

The Cooperative Constant

Return to the sandpile effect. A mathematical power law governs the probability of a given magnitude of collapse. This seems to work with other systems as well: species extinctions, stock market adjustments. Bigger events have a severely declining probability. Then, would a power law perhaps work in both directions? Might there be a counterpoise to the destruction of complex systems that quantifies the frequency and magnitude of emergence out of systems at the edge of chaos?

A different sort of power law was envisioned in Thomas Hobbes' conception of human life in the state of nature as "solitary, poor, nasty, brutish, and short"—where life was aggressively selfish, with competition strongly outweighing cooperation. For the sake of the preservation of the individual, Hobbes reasoned, personal power would best be transferred to some sovereign who would act for the benefit of society. After the initial act of the surrender of power (that proceeded out of fear), management of the system was strictly from the top down. Hierarchical integrity reigned for Hobbes. The sovereign was an authoritarian figure and typically ruled over a system that favored the wealthy and powerful class, with little or no consideration of democratic values or processes. To Hobbes, human beings lived in a dualistic and contentious world, as self versus other, rather than in a relational system, that is, as self with other.[17]

Alternatively, our thesis is that competition and strife is exaggerated in the long history of our species. Hobbes' attention, like that of many contemporary journalists, was drawn to human conflict and violence simply because it is so spectacularly shocking. Anything noisy, threatening, frightening, explosive, bloody, brutish, and revolting in the behavior of our species makes the news above all other aspects of our lives. In place of a predominantly competitive ethos, we highlight a hypothetical constructive principle with enormous staying power in the long term. Evolutionary history informs us that cooperation, writ large, has commonly prevailed from the beginning of our universe, leading to a vibrant world of complexity rather than a static and stultifying simplicity. We are heirs of this struggle between the Hobbesian view of the competitive-combative ethos in nature and society and the opposing evolutionary principle of free association and cooperation.

Thus we think it is evident that throughout biological and social evolution there has been a principle operating that we have called the *cooperative constant*. We adopt the term metaphorically, not in any strict mathematical sense. However, our usage refers to a strong trend having profound formative influence in prebiotic and abiotic nature as well—for example in various forms of chemical catalysis. Major examples in biology include multicellularity, endosymbiosis and exogenous mutualism, sex, and altruism. The initial encounters leading to such phenomena appear non-Darwinian (in the sense that self and other at any level of organization meet in noncompetitive congress). Only afterward does natural selection sort and sift the best fits between and among incipient cooperators. In pages to follow, we suggest that cooperation in universal evolution may have a persistent, or at least sporadic, slight edge over competition, maintaining significant hope for benign emergence in the future.

If a power law were to predict milestones in evolution, a possible trajectory would have complex catalytic polymers such as RNA and proteins emerge infrequently out of near chaos on the flanks of molecular attractors. Microbial life-forms would be far less likely, even in favorable conditions; integrated multicellular expressions scarcer still; self-awareness and technological manipulation of nature would be ever more rare, until, far out on the limb of the power curve, we might try to imagine some ne plus ultra of complex emergence in a universe of becoming.

Human hubris, however, is unwarranted, and our probable situation sobering; here in our time in the sun we may well share a galaxy of impossibly scattered anthills or, as cosmically isolated microbes, merely metabolize furiously our transient spot of resources. Only the real microbes in the rocky rind of the planet[18] have a very high survival potential. But they have been in stasis (morphologically, although not biochemically) for nearly four billion years.

Still, it is amazing where we stand as a consciousness in the cosmos. At last (religious fundamentalists, get thee behind!), we have a creation-based world.* Cultural evolution and memetic selection[19] have essentially replaced their organically derived counterparts. But there are never guarantees for any specific emergent entity. Avoiding our own attractors toward stasis and aggression may be the only guideline that permits us to keep playing in the cosmic surf. Our fate may be only to spawn new stuff and die. Nevertheless, we suggest, there is hope that comes out of the cooperative constant and even sensitive dependence of evolution on initial (or prevailing) conditions. We can never predict specific outcomes, but perhaps we can steer roughly and fractally toward greater participation in the universe.

---

*Creation is now rife in biology, with genetic manipulation and approaches to spawning new forms of life itself. And in a playful speculation on "philosophical implications" of new cosmological insights, the physicist Andrei Linde suggests the intriguing potential of creating whole new universes in the laboratory. See Linde's web page: http://www.stanford.edu/~alinde/.

PART I
_____

Origins

Primal Emergence

The universe looks like a huge growing fractal. It consists of many
inflating balls that produce new balls, which in turn produce more
new balls, ad infinitum. Therefore the evolution of the universe has no
end and may have no beginning.

—Andrei Linde, Stanford University

The instant it happened, cosmologists have surmised, an unfathomable,
dimensionless seed burst with energy that might sustain its growth
forever. This was the hypothetical Big Bang that gave rise to our
universe. Explosion is an imprecise description for what presumably
happened; inflation and expansion are somewhat better terms, for everything
was internal to the process. Cosmologists now usually reserve "inflation" for
an incredibly rapid size increase of our universe in the first moments of its
appearance. The initial event of the Bang is still—and probably always will
be—a theoretical construct, utterly impossible to verify by any observation or
measurement. However, physicists have fairly confidently backtracked—guided
by microcosmic experiments that accelerate subatomic matter and achieve enormous kinetic energies—to envision the macrocosmic environment inhabited by
such matter just trillionths of a second after the Big Bang. A carefully calculated
model of this earliest-simulated environment indicates that at this time the temperature was about 1,000,000,000,000,000 degrees (one quadrillion, or 10^{15} in
scientific notation). This calculation and the description of early cosmic development that follows are based on classic to recent interpretations by physicists
worldwide.[1]

At such a temperature, the universe we now inhabit must have been a bubble of
ultra-dense gas called a plasma and must have been expanding at an inconceivable rate. In fact, by the time the temperature had cooled to 10^{15} degrees, most of
the inflation was already over, and space extended through a vast volume. Even at
this very early stage, all of the kinds of subatomic particles familiar to nuclear
physicists, as well as their antiparticle counterparts, were in play: matter and antimatter coexisted briefly in enormous quantities.

Shortly thereafter, but still within the first minutely split second of universal
existence, most of the particles and antiparticles neutralized one another and
disappeared in an enormous, cumulative surge of energy. However, riding ahead

of this surge with an enormity of significance (in which we have written and you are reading this paragraph) was a tiny excess of matter over antimatter.[2] It appears that quarks outnumbered antiquarks and electrons exceeded antielectrons (positrons), among other such particle pairs, by about one in ten billion. Within the next few minutes, combinations of this primordial matter produced the nuclei (but not yet the complete atoms) of the lightest elements in our universe—mostly hydrogen and helium.

Modern cosmological science holds largely to the Standard Model of high-energy physics, which traces the origin of our universe to the Big Bang, although challenges to the Standard Model have appeared (one of them concerns the matter-antimatter ratio). Nevertheless, the Standard Model is well-reasoned, backed with mathematical rigor, and very sober stuff. Overall, the epic grandeur of the subject is unparalleled and has moved even some scientists, in descriptive and reflective writings, to passionate and romantic discourse. We hope to reflect some of the excitement of the progressive discovery of our universal origins in this chapter. Our inspiration is illuminated by the trajectories of giants. Carl Sagan's voice has now diffused through light years of space, and recent discoveries have begun to shine with flashes of passion in some of the writings of leading physicists and cosmologists in this field of tracing universal evolution—Steven Weinberg, Leonard Susskind, Neil Turok, and George Smoot, among others—on which we have based this sketch of ultimate beginnings.

The matter-antimatter ratio was clearly a lucky happenstance. Yet another astonishingly favorable "setting" of conditions at the earliest moments of the universe, according to the model, was a concentration of net vacuum energy, now often called dark energy (see discussion later in this chapter). This condition, discovered by Einstein, is also known by its more philosophical-sounding name: the cosmological constant. The vacuum of space is not the same as nothingness; it has never been empty, and remains filled with energy (and scattered motes of matter). That energy has been governed by a number of contributing forces that began with the Big Bang, and cosmologists have recognized that an almost incredible balance long existed between expansive forces that have inflated the universe and opposing forces, notably gravity, that operate to collapse it. Without the extreme fine-tuning of the cosmological constant, our universe would either have expanded so quickly that matter could never have coalesced into stars and galaxies, or it would have collapsed back on itself long ago. Either way, life would have been impossible.

Indeed, among leading theories of universal origins, and compatible with the Standard Model, is the idea that ours is only one of a near-infinite variety of universes—thus part of a collection of universes called the multiverse, or megaverse—with variants that have existed and will continue to bubble into existence throughout eternity. Conditions such as the cosmological constant that govern duration and distribution of energy and mass may be different with every Bang (some may be mere Pops), and our universe can be seen as one of an infinitesimal few capable of generating life.[3] With an eternity of Bangs, time stands outside our universe, and nature at large runs on an infinite clock.

The ancient Greek atomists, led by a progression of scholars, notably Leucippus, Democritus, and Epicurus, some three hundred to four hundred years before the Christian era, defended this view of an infinite and near-infinitely divisible and eternal void as the arena of existence. Thus they covered all bases for their universe, and filled it at the lowest limit with indivisible particles (atoms—the Greek word *atom* means uncuttable or indivisible). Epicurus, in particular, saw material objects in nature as associations of atoms subject to rearrangements and evolution; he considered even the gods and goddesses as material beings subject to the same random and non-teleological emergence of events. There was no universal purpose in and of itself for Epicurus. An infinite number of atoms, colliding into clusters, made up everything—they were the world, and if they were infinite, the possibility of infinite worlds also existed.[4] And, he reasoned, if atomic assemblages manifesting perceptible nature were broken up into individual atomic particles, reassembly would proceed spontaneously to form new objects in the universe.[5]

Competing philosophers such as Aristotle differed, arguing that there were no atoms and no spatial void. Below the moon, a continuous fabric of the four essential elements (*aitea*)—earth, air, fire, and water—composed material nature, which takes forms that are predestined. Objects or organisms may change, but they contain a sort of embryonic program that takes them toward a final form. First the material cause, the "from what" (*to ex hou*) of which things are made (ingredients), then the efficient cause, or the "by what" (*to hupo tinos*) by which things come to be (forces that generate), and then the formal cause, the "what" (*to ti esti*), the idea contained within anything that circumscribes its specific nature or identity, and lastly the final cause, the "for what" (*to hou heneka*), possibly presaging emergence or evolution, bespeaking the purpose, or destiny of the thing (the raison d'être).[6]

Whereas the atomists imagined the relationships of the various atoms and how everything in the world emerged from their atomic combinations, Aristotle saw the essence of things in their formal cause, that is, in some ontological structure. The acorn, for example, contained the ontological structure of the oak tree in its formal cause, given the soil, moisture, and sunlight provided by the efficient cause or forces acting on it. Acorns can become nothing but oak trees in this model, but as we will see later, in the Chinese worldview, as the comparative philosopher Roger T. Ames has wittily suggested, 99 percent of acorns never reach their purely oak-tree destiny, for they become meals for squirrels.

Meanwhile, his thinking on connected essences led Aristotle to the conclusion that the universe consisted of concentric spheres all the way out to the stars. Above the moon, all objects were composed of a fifth element he called quintessence. In the outermost sphere, Aristotle speculated the existence of a Prime Mover that represented, along with the motions of the stars, an eternal, constant perfection. On one level, Aristotle's Prime Mover was a vision of God completely divorced from its creation; however, on another level, it is the highest reality or being as pure potential. Without this potential presence, the material universe would have gone nowhere, just as the acorn in its microcosmic way must be guided by its essential potential to become an oak tree. Thus his God represents not so much the beginning cause but the final cause toward which the universe strives for perfection. This God does not act, did not materially create the world, for it is co-eternal with it, and is only logically capable of merely contemplating itself forever in abject isolation.[7]

Much later, Christians such as Augustine and Thomas Aquinas adapted Aristotle's views to bolster biblical notions of the finiteness of time and space, envisioning God as the ultimate authoritarian, a creator who stood outside the universe and had put it on a definite schedule.

Recombination: Then There Was Light

To biblical literalists and, indeed, many people outside of the cosmological community, it may seem surprising that our early universe, seething with radiation and packed with flying electrons and particles of initially condensed nuclear matter, developed in darkness or at least dimness. The "shining" of early photons (the common units of electromagnetic energy) was mostly far beyond the gentle radiant manifestation we now classify as the visible spectrum. Even as cooling progressed, photons nearly universally irradiated space with gamma and X-ray wavelengths and energies for hundreds of thousands of years. And in the primordial cloud of plasmic-matter, photons were hardly free to glow at any wavelength before they collided with nearby masses, especially electrons, and were reabsorbed, then reemitted briefly, absorbed again, and so forth. Matter restrained light. Then, at last, the universe suddenly became transparent, and thereafter gradually cooled through the spectrum to reveal, ever more widely, the wavelengths of the rainbow.

What happened to brighten the universe was that everything was becoming "cool" enough for the plasma-gas of photons, electrons, and lightweight atomic nuclei to radically change its character—a transition called recombination. The nuclei began to capture and confine many of the photon-retarding free electrons, and now true atoms formed for the first time. Electrons are naturally attracted to nuclei, owing to their opposite electrical charges, but this relatively weak attraction could not happen until temperature and kinetic energy declined to a critical threshold. However, as the electrons joined nuclei, their energy dropped suddenly to a new low level. The abandoned energy of the electrons appeared as new photons, and now the photons could travel much farther without interference. All of space had cleared like the air after a rain shower, and its radiation, now mostly unbound, illuminated the scene much more widely. At this stage the temperature everywhere was something like that near the surface of the sun, and very nearly uniform, but with surprising small variations here and there, like warmer and cooler spots in an early-summer swimming hole. Poised to evolve in complexity and promise, the universe entered a new, quickening season.

Remember, as yet there were no stars or galaxies; matter had become organized merely into the simplest of atoms whizzing through the void. But at this critical phase of universal evolution, the forces and energies that now filled space and time relaxed into patterns that would prove discoverable and even logical (if not always intuitive). This threshold arrived some 380,000 years after the Bang,[8] and these discoveries would be made much later by us, as living derivations of those atoms, after they had been arranged and energized ever more gently through emergent passages of an incredible journey. The evolution of this universe ultimately led to an awareness of itself.

Uniformity to Pattern: A Cosmic Butterfly Takes Wing

In theory, the Big Bang itself began as an "explosion" of absolute symmetry, starting from a virtually infinitesimal point. At this formative scale of universal (or pre-universal) existence, the only nonhomogeneous effects were those that physicists calculate and conceptualize as random quantum fluctuations. Those fluctuations occur generally in energy fields at subatomic levels and have persisted throughout the history of the universe. When the Big Bang happened, during its inflation phase quantum fluctuations in the primordial energy field expanded with space virtually at once, stretching beyond the scale of galaxies. Cosmologists generally agree that this enlargement of quantum turbulence took place in less than 10^{-33} of a second. Close to the very beginning of everything, a roughness was imposed on the universe. Its properties—energy, mass, and the forces uniting mass-energy—began in a completely unified state, but they would not stay that way for long and would be profoundly affected by the stretching of quantum irregularities. Quickly, the smooth homogeneity of everything developed waves, lumps, and wrinkles, especially variations of the density of energy and mass that began to structure space and time. Also, superimposed on the enormously magnified quantum effects to further shape the cosmos was a process physicists have termed *symmetry-breaking*.

The small surplus of matter over antimatter was only one of the asymmetries. Equally profound, engendering structure out of the matter that remained, was gravitational energy that broke out of the unified energy field at the beginning of the universe we inhabit. Following inflation, gravity amplified its effects throughout space in response to the stretched quantum fluctuations that first set the patterns into which structure would evolve. Matter began to concentrate in some regions, leaving other areas relatively less dense. The distribution of galaxies would later correlate with this initial pattern of lumpiness. In other ways, the early shaping of our universe may have progressed through discontinuities emerging out of symmetrical force fields that then took particular forms within the wrinkled "quantum fabric" of spacetime.[9] One seminal example that led out of physics to chemistry and, ultimately, biology was a unified particle symmetry that concerned an electron-neutrino unity. These particles assume a smooth, uniform identity—virtually pure energy at an extremely high temperature—but, upon cooling to a certain threshold, suddenly break into unique entities, with the electron assuming much more of its energy as mass and the potential to build the emergent complexity of chemistry around elemental matter. Thus, the early breaking of symmetries led to various subsequent processes to shape the universe on all scales with an inexorable potential for the emergence of everything.

Symmetry-breaking that generated primordial complexity in the early, rapidly cooling universe has been compared to the transitions many common substances go through as they change physical character from gas to liquid to solid compositions. Technically, such shifts of organization in matter and energy are called phase changes. Perhaps the most familiar example is water. Its changes in structure and symmetry—from vapor to liquid at cooling, to vapor at the boiling point, and from liquid to ice at the freezing point—crudely parallel some of the shifts in symmetry and coalescence of structure in the early universe.

To modern students of complexity, such phase transitions represent the potent state of being known as the edge of chaos (see introduction). The structural organization and symmetry of water change swiftly and radically at the temperatures of both boiling and freezing. Condensing from vapor to liquid, water molecules begin to stick together in clumps, albeit jostling and oscillating with variable localized distributions of energy, breaking free and rejoining liquid patches of all shapes and sizes. The symmetry of the gas—smoothly, randomly similar in every direction, is broken into a more constrained but locally chaotic pattern of interacting molecules forming the clumps and clusters of the liquid. At the freezing point another dynamic transition ensues, and the symmetry breaks again from the clumpy-chaotic to the more highly constrained crystalline. The point to be taken from such models, even the simplest, as represented by water, is that the systems of nature reveal seminal nodes—tipping points regarding symmetry or structure or behavior or all of them combined—where suddenly everything changes; new properties spring forth; novel worlds take shape.

Some philosophers, hearkening back as far as Aristotle as we saw above, and even a few physicists have toyed with an idea called the anthropic principle: that there is some Prime Cause, or directed, mystical quality inherent in our universe. The universal "settings" are all so perfectly in our favor.[10] Did our universe anticipate us, or something like us, that would try to understand its story and decipher its meaning? Most cosmologists resist the anthropic principle; the multiverse (infinite Bangs) extension of the inflation hypothesis gets us off the hook by suggesting that among all the universe start-ups, at least one like ours was inevitable. We are here simply because we cannot be anywhere else.

Instead of some top-down, cosmic directorship, ours looks to be a place of serendipity that somehow leads to marvelously emergent self-organization. It began with absolute simplicity and the most basic physical laws, then improvised and leveraged increasingly sophisticated "cause" along the way. Among the rules, and perhaps the farthest-reaching of them, is sensitive dependence (see introduction). But this could not be dependence on initial conditions in the case of the Big Bang. The first sensitive dependence would have been in response to the initial disturbance, quantum fluctuations as asymmetry in the inflating universe. Some of the latest interpretations of cosmologists point to an amazing butterfly effect—the biggest and most persistent in all time and space that we can know.

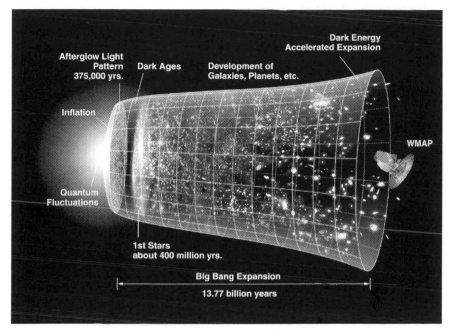

Figure 1.1 A depiction of universal evolution. The Big Bang begins at the far left, followed by immediate inflation. From recombination (here termed "afterglow"), when radiation (the universal photon field) was released from plasmic confinement, the universe gradually expanded as stars and galaxies formed and reformed until fairly recently, when dark energy began to increase relative to gravity, and space started to grow more rapidly again. *Credit:* NASA / WMAP Science Team.

It presumably depended on the early condition of asymmetries; then it unfolded directly from disturbances in the gas density of the primordial plasma during the momentary interval of inflation.

The Music of the Plasma

Although the universe was opaque, or nearly so, until the phase of recombination, it was anything but silent. Echoing ancient Vedic religious thinkers, who speculated that "om" is the universal intonation, contemporary scientists realized the universe began to sing very shortly after its birth and continued with a deepening voice for about 380,000 years—that is, until the time of recombination.[11] Although not considered especially significant until the 1990s, this realization first surfaced in the 1960s when researchers in the United States and the Soviet Union proposed that sound was a property of the plasma that filled the early universe. The sound of the universe began with kinetic energy playing through small variations in plasma density: compression and rarefaction. Owing to the almost instantaneous inflation after the Big Bang, those density variations (sound waves) of the same intensity appeared on all scales virtually at once. This started all the sound waves in phase with each other. The universe had tuned itself like a musical instrument, and it began to play with overtones or harmonics, like an organ in a cathedral.

Compression in a medium raises its temperature, and rarefaction induces cooling. Thus the distribution of temperature through the plasma patterned itself exactly in synchrony with the sound waves. Slightly warmer and cooler regions formed in space, marking extremes of high and low plasma density. As the universal song extended through time, these regions expanded, and they oscillated in temperature with the beat. Then the song of the plasma ended at the recombination phase of the universe, when matter abruptly thinned and stopped the music. However all of the sounds then playing left a ghostly imprint on the cosmos in the shape of final warm and cool patches in the radiation (photon) field of space. The patches were of varying sizes, depending on each particular acoustic wavelength in the pattern of overtones that existed when the song stopped playing. The largest patches of all were cool and represented the traces of the fundamental wave of the universal sounding board—stretching from earliest compression to final rarefaction at the time of recombination.

From this time, the immortal song was recorded in *light-temperature* (the energy of photons), and even galaxies would eventually dance to its rhythms. This is a milestone in modern cosmological understanding of the universe in its infancy, elucidated during three decades of brilliant research involving numerous scientists from many countries.

The rest of this story takes us through spacetime from the abrupt phase of universal recombination, with its release of wide-wandering photons, down to scientists' observations of those photons, now greatly cooled to around 2.7 degrees on the absolute (Kelvin) thermometer. When these ancient photons were discovered in 1965,[12] coming at us from all directions of space, they were cumulatively dubbed the cosmic microwave background radiation, or CMB. Very soon, cosmologists realized they were detecting the faint afterglow of the hot early universe. Following the Standard (Big Bang) Model, they then realized that the CMB specifically traced to the time of recombination and were able to calculate the approximate date of its initial release—380,000 years after the Bang. At first the glow seemed absolutely uniform in temperature. However, in the early 1990s, using increasingly sensitive equipment, scientists first confirmed slight variations in the temperature of the CMB.[13] Those variations, now carefully mapped across the visible cosmos, show up as the different-sized patches that are warmer or cooler than the average reaches of space around them.

Thus when the music of the plasma died with the recombination of matter, its final song was imprinted in those virtually uninhibited photons that spread from this significant cosmic horizon. Photons emerging from the compressed regions of sound waves were slightly warmer, and those from rarefactions slightly cooler than the average. Now we can trace them back to discover the harmony that prevailed as the universe went into an especially pregnant phase change.

The astonishing finale of this story to date is that out of the song that recorded itself across the cosmos in radiant immortality, there also emerged the universe's dominant physical structure. Over eons, the slightly elevated gravity of compressed (denser) regions pulled matter together into patches of stars, quasars, and galaxies, while the rarefied zones remained relatively empty voids. Scientists mapping space have now confirmed these patterns on scales up to more than several billion light-years.

Dark Enigmas

The classic American philosopher Charles Peirce conceived of nature as initially in a state of randomness, but with an inherent, albeit weak, property to take on "habits."[14] Out of this property, he envisioned a self-organizing tendency would ensue from which large-scale regular patterns could evolve from very slight shifts in formative conditions. His views seem to anticipate trends of universal self-organization such as those that were seeded at the time of recombination and recorded in the scattering of the CMB. Moreover, Peirce argued that the universe might change even its most basic habits, subjecting science to wrenching revisions of entrenched interpretations of nature.

Cosmologists have known for some time that, in our universe, what we see is not all we have. In fact, ordinary matter that appears as quasars, stars, planets, nebulae, galaxies, and other visible substance seems to constitute just a very small fraction.[15] Current theory holds that around 4 percent of the universe is made up of normal matter—the stuff of stars, planets and people—and somewhat over 20 percent of it is dark matter. Dark matter does not interact with ordinary matter or with visible light shining through space, except by contributing to gravity. Subtle observations by astronomers indicate that this dark matter is distributed in clumps and clusters arranged in close proximity to normal galaxies. So, it too has been organized by the primal forces emerging from the Big Bang. However, just what makes up dark matter and whether we would know if it were slipping through our fingers remains a mystery.

This leaves nearly 75 percent of the universe that is thought to have a very exotic composition indeed. That remaining component, more dramatic than dark matter (and unrelated in any specific sense), is dark energy, a recent term in science for what passes in science fiction as antigravity. This concept actually was part of Einstein's Theory of General Relativity, but until 1998 physicists had largely ignored it as a mathematical artifact and not part of our tangible universe. In that year came the first observational evidence that the expansion of the universe is now accelerating, rather than continuing to decelerate after its earliest phase of inflation.[16] The observations were made, by the Hubble Space Telescope, of relative brightness coming from supernovas—brilliantly exploding stars—of a standard type with a well-known luminosity at precisely calibrated distances from Earth. Intensities of light detected from a series of these supernovas (as related to their distances) did not fall on a smooth curve. The farther explosions appear relatively bright considering their distance (their photons started toward us when the expansion of everything was slowing down, so the distance traveled was comparatively short). But the nearer supernova emissions were dimmer than expected, meaning a longer travel path than predicted (other possible explanations, such as intergalactic dust, had been reliably ruled out). The break point in the relative luminosity curve appears, for these supernovas, to occur at a distance of around five billion light-years. The light has taken that long to reach us—so we are seeing a change that began about five billion years ago. Since then, like fireflies being scattered by a rising wind, the substance of the universe has been dispersing ever more quickly.

In one imaginative effort to explain the apparent antigravity property of dark energy, New York University physicist Georgi Dvali proposed that normal gravitational energy in the form of its hypothetical quantum units, called gravitons, is leaking out of our observable universe.[17] The gravitons disappear into multidimensional space that could lie undetected around us (see next section). So far dark energy's relative density is such that it only affects matter on the largest scales. It overcomes gravity to the degree that perhaps large galactic aggregations can no longer form and indeed may be starting to come apart. However, subtle indications in the CMB suggest that dark energy has increased with time and now rules over all other forms of energy and matter. If it continues to grow in strength, cosmologists predict it could take apart galaxies themselves, and eventually all organized matter down to atomic nuclei. Dark energy would be the ultimate attractor of dissipation.

Star Struck: The Enabling of Life

Following the recombination phase of our universe, the first stars began to form by gravitational attraction. Everywhere responding to gravity, they coalesced into galaxies. The earliest stars were made almost entirely of hydrogen and helium, with a pinch of lithium. Those three are the lightest and simplest of the pure chemical substances known as elements and were first formed at recombination. As yet there were no heavier elements, but they would soon appear in the interiors of stars.

Stars are pressure cookers for creating elements heavier than lithium.[18] As beginning science students learn, the weight of an element refers to the weight of its atoms, and is nearly all in the nucleus, made of relatively heavy subatomic particles called protons and neutrons. (Remember those particles are thought to be composed of even more basic building blocks known as quarks). Protons are positively charged, and their number defines a given element—hydrogen has one proton; helium, two; lithium, three, and so forth. Neutrons can vary in number within a small range for each element; for example, the most common form of hydrogen has only the proton as its nucleus, but another form has a proton and a neutron. Such a variant is called an isotope—the "heavy hydrogen" isotope just mentioned is named deuterium.

Note that in the ultra-hot, dense milieu in the interiors of stars (as in the very early universe), complete atoms do not exist. Their final components, the lightweight and negatively charged electrons, remain free—too energetic to be captured by the nuclei. Only in cooler realms will electrons be attracted by protons to surround the nucleus in a structured cloud. Then, in pure elements, at relatively modest temperatures, the electrons match protons in number—making for electrically neutral atoms.

All the precise recipes for cooking up heavy atomic nuclei deep in the interiors of stars are not necessary to state here. Suffice it to say that the enormous temperatures and pressures within the stellar furnaces force particular combinations of hydrogen, helium, and lithium nuclei to come together. Then the products of those fusions, such as carbon, combine with sundry others to form yet heavier nuclei—all the way to iron—in thermonuclear reactions that release immense energies. Lighter elements are thus "burned" in stars and the "ashes" emerge as brand-new heavier elements. The energy released by this burning is the energy of suns.

Carbon, the matrix element of life, has one of the most thrilling creation stories in our universe of complexity.[19] Deep inside a star, two helium nuclei are crushed together, momentarily forming a heavy and unstable nucleus of beryllium with four protons and four neutrons. Too many neutrons makes beryllium unstable. The tendency of this particular heavy isotope is to break apart (fission) almost instantly, reverting to its original components. But in a substantial number of cases, beryllium fuses with another helium nucleus creating carbon (six protons, six neutrons) known as carbon-12 for the total makeup of its nucleus. At the moment of its formation, carbon-12 exists in what nuclear physicists call an excited (high-energy) state. But unlike the heavy beryllium, it sheds excess energy not by fissioning, but simply by emitting a photon (quantum, or ray of pure energy). Owing to size and spatial packing of its protons and neutrons, this carbon nucleus then settles into a configuration that is virtually fission-proof.

However, this critical process of atomic creation depends on an almost incredible precision in the energy of the excited state as carbon-12 is formed. The excited state, in turn, depends on the energies of the beryllium and helium nuclei as they fuse—if slightly too low, fusion would fail. However, if carbon's excited state were slightly higher on the energy curve, in the infinitesimally longer embrace to achieve fusion, the beryllium would fission apart before any significant amount of carbon could appear. This exceedingly fine adjustment in the energy of carbon formation is another of the universal settings (albeit belated) in our favor. In the hot, plasmic, stellar depths, carbon's nucleus itself becomes a new building block—leading to oxygen, nitrogen, calcium, and other essentials of life. Without ample carbon in our universe, we would not be here to look back along our ancient pathway of becoming the universe.

Cool Runnings

Following stellar synthesis of all the elements crucial to life—and more—supernovas spread them far and wide. Supernovas are colossal explosions of certain relatively large stars at the end of their existence. Such stars reach an abrupt crisis in old age. Cycles of heavy-nucleus building by fusion have almost ended; there is little lightweight "fuel" left, and the dying fusion energy that creates an outward pressure from inside the star can no longer support its bulk against the unbelievable gravity that is generated by its mass. So the outer layers, now rich in carbon, oxygen, nitrogen, silicon, and so on collapse toward the last-formed and central core of iron, and the resulting rebound is the explosion, with the energy of up to a billion suns. Once away from the stellar furnace, atomic nuclei can attract and hold electrons, becoming complete atoms in the process. All that star stuff, now with the potential to become our stuff, is scattered in nebular dust clouds through light-years of space—until it comes together again in a new cycle of gravitational attraction and star formation.

Sometimes such renewed attractions produce planets, with generous shares of the heavier elements, around their second- or third-generation stars. And on and around the much cooler environs of planets, elements can behave in vastly more complex and intricate ways. Chemistry now emerges far beyond the intensely hot, brutal bashing of atomic nuclei, for in its dominant workings, chemistry resides in the electrons and especially in the rarefied realm of electrons in the outermost parts of the stratified cloud they occupy, well away from the nucleus.

At the mild temperatures that characterize planets, even in a volcanic eruption or exploding dynamite, most atoms are whole. In each, a nucleus at the center maintains, by electrical attraction, a kind of structured "atmosphere" of electrons that becomes more complex in heavier elements. Again, the precise details are not needed here for an overview of the lay of the emerging landscape. Those details can be found in any basic high school chemistry text. What is important is that as the milieu becomes cooler, chemistry builds complexity to produce some of the most intricate and interesting things that appear in the universe—but not where it gets too cool. Like Goldilocks' preferred porridge, the ideal condition lies in a mid-range of temperature. Most parts of the universe are now very cool indeed, and matter and energy dwell there in relative simplicity.

It is perhaps surprising that interactions focused merely in the electron-cloud tops of atoms ultimately took the evolutionary lead in building molecules that, depending largely on size and temperature, phase themselves into the gases, liquids, and solids of everyday substances. Among the molecules, those constructed with a framework of carbon atoms proved to be especially versatile. As noted above, carbon as a naked nucleus burning in stars was crucial in generating other essential basic ingredients of an (explosively) emergent chemistry. But in its second coming, with an electron halo, abundantly manifest in condensing clouds of supernova ash, carbon exhibits an especially cooperative (covalent) tendency to attract other atoms. It joins readily with its own kind, forming carbon chains and ring clusters, as well as with an ecumenical range of atoms of other elements, especially hydrogen, oxygen, nitrogen, phosphorus, and sulfur. Constructive carbon builds the framework for an almost infinite variety of organic molecules, each with unique structure and properties. And some of these materials then assume a potential to generate the amazing dynamic complexity and ever-more-potent capacity to cooperate that catalyzed the most profound emergence of all.

Physicist Andrei Linde's characterization of the universe quoted in the epigraph that opens this chapter: "the universe looks like a huge growing fractal . . ." refers to the multiverse model of serial inflations. But in the particular universe we now inhabit the fractal organization of nature has progressed through ever more refined forms for 13.8 billion years. The human mind and human society lately have become parts of this swirling, diversifying, branching, evolving, and emerging continuum in which the fractal self is now at work and in play. In this chapter on physical origins we have glimpsed touchstones of the cooperative cosmos uncovered by science, tracing the long road to our time of complexity in universal history. In Chapter 2 we will focus on microcosmic origins of the fractal self as an emergent human spirit in our ancestors seeking connections in the grand milieu of nature.

Out of the Dreamtime

Without myth every culture loses its healthy, creative natural power.
—Nietzsche, *Birth of Tragedy and the Genealogy of Morals*

I n chapter 1, we traced the beginnings of evolution in our universe as shaped
by universal physical forces. The Big Bang and its aftermath might be called
the "creation story" according to science, in which tremendously wide-
ranging order and deterministic pattern propagated from a cosmic seed of
perfect uniformity and smoothness. We saw how formative properties of matter
and energy were forged through turbulence and an emergent principle of attraction that seems to pervade all of nature. This fractally structured emergence
subsequently enabled development of the cosmos' complex forms and behaviors
in ways that we are just beginning to understand. Complexity in the cosmos organized itself from the bottom up and built, across scale from nanometers to
parsecs and through billions of years, worlds so wondrous that they intersect
with dreams.

As it emerged out of simplicity, the universe adopted a modus operandi
that we are calling the cooperative constant, initially manifested in physical
forces, especially gravity, and progressively complemented by chemistry (see
also chapter 3) in which various elements developed bonding attractions, building
complex patterns of structure and generating chemical reactivity. Ultimately,
catalysts that intensified "intimate" molecular connections would evolve, weaving their influence within ever more advanced biochemical manifestations of
the universe to quicken the coming of organic complexity. From an evolutionary point of view, an emergent catalytic potential, an attraction to cooperate, or
participate in heterogeneity—which becomes a sine qua non for the existence
of life—is widely characteristic of matter and is now found at the heart of the
most progressive systems of which we are aware.

In this chapter, however, we diverge from the scientific pathway to follow some
primordial strands of thought on origins. The thesis of an intimate, participatory universe appears to have been anticipated well before human beings began
to articulate worldviews in formal philosophic or scientific terms. Here we begin our search for the fractal self by tracing what appear to us to be prototypes
recorded in myths and oral histories, reflecting perhaps unconsciously the
yearnings of early people to participate in universal evolution.

It's not a surprise that these ideas constitute foundations of our cultural understandings of the position we occupy and roles we play in the world around us. Children in their first outdoor explorations commonly begin to think of themselves as embedded in nature, and, as a species, we passed through a sapient childhood. Our first self-reflections and meditations came perhaps in considering the notion of human selfness as intimately connected to the natural world. A universe of apparent indifference (eventually interpreted in Darwinian terms of struggle for existence) was most likely a later construct. Without philosophy and science in those early times, the only recourse—to explain why something instead of nothing and where the incipient self belonged in the world—lay in myth, the art of telling intimate stories of origins. And the apparent posture of participation expressed in the oral mythical recitations of disparate societies seems to bespeak an ancient holistic grasp and appreciation of humans closely integrated with nature, amid creative turbulence and emergence. The fractal self appears deeply rooted in our earliest evolved psyche.

At Home in the Participatory Universe

Many cultures developed creation (or cosmogonic) myths that spoke of the forging of the Earth and universe by various gods and goddesses—some kindhearted and others of the crueler sort. We were polytheistic by nature (an omnipotent, authoritarian God was a later invention) in our attempt to explain the sheer existence of so many things in the world. Our formative gods and goddesses represented our multiple aspirations and concerns and our widely pervasive sense of self. For the early tribal Greeks, the Homeric self was constituted as a multiplicity of centers acting as receiving stations for the various divinities to work through. The Homeric *psyche* (the word later used by Greeks to represent the unity of an individuated self), or soul, was constituted by an openness to the personified forces of the natural world. In a profound sense, we were originally *in* the world; we did not view ourselves as being distinct, disparate, and separated from the very forces that make life possible. As we see in Homer's *Iliad* and *Odyssey*, the parts of the psyche—the *thumos, kradie, etor, ker,* and *phrenes*—seem to be personified receiving centers for the gods and goddesses.[1] Our gods were channels to the many world-lines we encountered and considered through a lifetime. Our original myths across the planet would bespeak a self that was defined in relation to natural phenomena. This self's original being was posited by the nineteenth-century French philosopher, ethnologist, and theoretical anthropologist of non-Western mythology Lucien Lévy-Bruhl as one of "mystic participation," in his 1926 book *Les Fonctions Mentales dans les Sociétés Inférieures* (*How Natives Think*). Commenting on Lévy-Bruhl's work, Owen Barfield describes participation as "the extra-sensory relation between man and the phenomena."[2] Lévy-Bruhl's work on human connectivity with nature has direct reference to our arguments for a fractal self, a self that is defined by its relation to the unfolding natural world.

Lévy-Bruhl understood the *participation mystique* as a way of being in the world that constitutes a perception of a self that is as much at home in nonhuman nature as in the social construction of its communities. Even today we can see how certain non-Western cultural conceptions of a self provide us with a means to backtrack into this so called "archaic" way of being in the world. As we noted briefly in the introduction, an excellent model for this excursion is found in the major insight of Thomas P. Kasulis. Kasulis focuses on a spectrum of cultural behavior and social behavior extending between polar manifestations of *integrity*, with its emphasis on individuality, and *intimacy*, which seeks cooperative relationships as the natural way of living.

Simply stated (and we'll return more deeply to this distinction in chapter 8), the Western self of integrity thinks of the natural world as being external, and consequently our relationship to the world keeps us apart from nature. The integrity self wholeheartedly believes the world can somehow be managed through our knowledge of it. In many ways, this sense of management explains the dominance of and superiority traditionally given to science in the West in contradistinction to other modes of knowing such as the mythic or poetic. Intimacy-based cultures, on the other hand, view the self as deeply embedded within the world, interlinked as a participant in its systems and open to mutual opportunities for change and emergence. For the self of intimacy, knowledge then resides at the interface between self and world with the emphasis struck on the internal relation of self and world.

The self of intimacy and its corresponding sense of mutualistic existence, explored by Lévy-Bruhl and other nineteenth-century anthropologists such as R. H. Codrington in the cultures of the South Pacific, can be traced all over the Earth. From his work as a missionary trying to understand Melanesian religious practice, Codrington became aware of a very powerful concept that lent an all-inclusive sense of being in the world. To Oceanic people the concept was known as *mana*.

> The word is common, I believe, to the whole Pacific. . . . I think I know what our people mean by it, and that meaning seems to me to cover all that I hear about it elsewhere. It is a power or influence, not physical, and in a way supernatural, but it shows itself in physical force, or in any kind of power or excellence which a man possesses. This Mana is not fixed in anything, and can be conveyed in almost anything; but spirits, whether disembodied souls or supernatural beings, have it and can impart it; and it essentially belongs to personal beings to originate it, though it may act through the medium of water, or a stone, or a bone. All Melanesian religion consists, in fact, in getting this Mana for oneself, or getting it used for one's benefit—all religion, that is, as far as religious practices go, prayers and sacrifices.[3]

In later studies, anthropologists appropriated the term *mana* to mean something along the lines of soul energy and a force that gives objects or persons standing, power, or excellence in the world; the term is still found in Melanesian, Tahitian, Hawaiian, and other languages of Oceania. *Mana* courses through all things, from rocks to the gods and goddesses in the great continuum, across scale and from the material to the spiritual.

Envisioning all things in the world and beyond as being infused with the life force of *mana* imparts a kind of fractal continuity to the diversity of all natural objects (rocks even have spiritual standing in this worldview) and living species from worms and snails and fishes to the supernatural (gods and goddesses). All participate in a wide web of being, joined by *mana,* or its cultural equivalent, as a sort of universal ether. The inherent divinity of all things is realized and celebrated.

A good example of this way of *"mana* thinking"—perhaps stemming from a very early antecedent concept—is found in Australia. After living ten years with indigenous Australians on an island off the coast of Tasmania, Robert Lawlor writes with sensitivity in his *Voices of the First Day: Awakening in the Aboriginal Dreamtime:*

> The Australian Aborigines speak of *jiva* or *guruwari,* a *seed power* deposited in the earth. In the Aboriginal world view, every meaningful activity, event, or life process that occurs at a particular place leaves behind a vibrational residue in the earth, as plants leave an image of themselves as seeds. The shape of the land—its mountains, rocks, riverbeds, and waterholes—and its unseen vibrations echo the events that brought that place into creation. Everything in the natural world is a symbolic footprint of the metaphysical beings whose actions created our world. As with a seed, the potency of an earthly location is wedded to the memory of its origin.[4]

In Lawlor's interpretation we see a kind of eerie poetic resonance with the Big Bang and the inflationary cosmos. For the Australian Aborigine, it is the potency for emergence that extends across scale at both the individual species level—the plant's image is found in its seed that will over time give rise to its recursive life history, being born time and time again— and at the universal level, where the memory of creation is found in the amplification of all subsequent form and process, and the "shape of the land" emerges in its complexity. Perhaps unconsciously evoking the edge of chaos, with its nearness to emergence, Australian Aborigines refer to this extraordinary potency as "dreamtime" in order to provide the temporal world with an eternal sense of sacredness.

Playing on the Edge: Realm of the Demigods

After the world is created, some form of tension or turbulence is needed to promote emergence in the evolutionary process. In the myths of many cultures, this need is satisfied by trickster demigods. Tricksters are semidivine shape-shifters and self-shifters; they flirt with chaos and dwell at the interface of the natural and supernatural worlds. In particular it is their role to push their mythic systems out of the initial frozen realms of elementary materials and forces. Tricksters are catalysts and, at their best, facilitators of the integration of human beings with nature as they shake up the world and move things into much more potentially creative patterns, fostering emergence and evolutionary change.

Coyote, a supernatural manifestation of that resourceful North American animal, is a trickster frequently associated with the notion of chaos in Navajo culture. Lurking at the creation of all things, Coyote followed the Holy People, the prime Navajo deities, as they carefully placed the sun and moon in the sky and began to sow the stars through the heavens in a static, orderly way. Quick as his namesake, Coyote yanked the blanket on which most of the stars still lay awaiting placement—upsetting the Holy People and scattering the stars far and wide across the sky as we see them today.[5] Later, Coyote was sometimes seen as an inimical figure causing calamities in orderly society, but in his primordial guise as the catalyst of constellations and intriguing star patterns, he appears as an agent of creative turbulence.

The Hawaiian hybrid deity Māui (who is also found throughout Polynesia) is another trickster (and perhaps more of a cooperator than Coyote). Māui's often precipitate actions move the world to the level of turbulence necessary for the creation of a novel, and often beneficial, state of affairs. Younger and smarter than his siblings, Māui possesses magical powers and a rogue-like nature. Although not a great fisherman in comparison to his brothers, who want nothing to do with him because he is so mischievous, Māui was ultimately able to hook the islands of the Hawaiian archipelago and bring them to the surface.

As a demigod, Māui is able to take advantage of both natural and supernatural situations in his quests to execute his schemes, which often proved useful for humans. Being able to change himself into different animal forms and blend into physical landscapes amid the volcanic terrain of oceanic islands, Māui's protean-like being is a forerunner to the shaman (also, in cultural lore, able to assume forms of various animals—see further discussion below). Such powers allow Māui not only to be adaptable, but also to foster opportunities for emergence. In perhaps his most celebrated feat, from his cryptic vantage point of surrounding cinder cones in Haleakalā (the giant crater Hawaiians named the House of the Sun) he lassoed the sun's rays as day broke over the great crater. Māui quickly convinced the captured sun to slow down its formerly rapid pace across the sky so more crops could be planted and harvested. The sun agreed to a compromise: hence, days in the summer are long, allowing for prosperous growing seasons.

It is not unusual that demigods such as Māui offer the gift of fire to humans. This gift, typically stolen or smuggled from the realm of the gods by the trickster, almost always causes a tumultuous upset for various orderly deities (this is the case in Greek myth as well as Polynesian). Such occasions, moreover, always seem to provide opportunities for the cultural evolution of humans. Māui and Hermes, his mercurial counterpart in Greece, both discovered how to make fire by rubbing dry sticks together. For Māui, however, it took a series of tries that placed him in the ironic position of being a victim of trickery himself. In one variant of the story, the ancestral mud hens, themselves demigods of a damp and misty realm, held the secret of fire. They were approached by Māui, who sought their instruction. The mud hens watched intently as Māui attempted to replicate their technique. On his third futile try, Māui threatened to strangle one of the mud hens in his frustration and suspicion that they were concealing part of the secret. Then it was revealed to him that the hard, dry sticks of the sandalwood, not the taro stalk or *ti* stem, must be used to kindle fire. For the trickery of the ancestral mud hen, Māui rubbed a red streak on her head, but showed compassion and let her go free. Māui was a god who could bring about emergent change in his world by either "craft or force."[6]

In ancient Greece, Hermes (Mercury to the Romans) was celebrated with epithets of the Contriver, the Wily, Shifty, and the Many-Turning as well as the Bringer of Luck, Ready Helper, and the Keen-Sighted and Watchful One. Hermes, as a trickster, made things happen in an otherwise relatively static world and, like Māui, sought the art of fire at a similar technological level: in Western mythology he is said to be responsible for inventing fire sticks. Although a player of pranks, Hermes is loved by many of the gods for his beneficial deeds, which are also extended to the human realm. As a benefactor of humankind, he guards domesticated flocks, guides travels, presides over business affairs, and serves as inspiration for harmonic music and speech—making him a sort of wide-ranging patron of the liberal arts. And with deft applications of his legendary swiftness in action, the many turnings of his travels and endeavors, and his lucky, speculative nature, Hermes dwells amid turbulence but manages to tame it in emergent order and pattern and creativity. Hence, in his brilliance, he becomes the god of communication, the Hermes *Logios,* the forger of orderly process in the world.

Hermes' integrating and order-generating role is born from his flair for transforming what might be chaotic events into meaningful change. But of course Hermes is more famous for his mercurial side as the "god of games and chance," another of his many claims to fame. His oscillating nature, between logic and order on one side and randomness and risk on the other, seems to frequently place Hermes in action along the border regions of attractors that are important to human beings. Creating opportunities for emergence positioned Hermes as an archetype for many of the Pre-Socratic philosophers such as Herakleitos, who stated a "thunderbolt steers all things" and that "for all things, Fire having come unexpectedly upon will pick them out and seize them."[7]

Divine Struggle: Between Intimacy and Integrity

Stormy occasions for creatively opportunistic developments find further expression in the ancient Greek demigod Dionysos, who was fathered by Zeus in dalliance with a mortal, Semele. The conception of Dionysos occurred in a kind of mythological Big Bang, as Semele was reduced to a pile of burning embers after she persuaded Zeus to show her all the glory his divine wife Hera got to experience. From the burning ashes, Zeus realized Semele had been pregnant and sewed the fetus of Dionysos into his thigh. Following such gestational chaos, upon the baby's birth Zeus asked Hermes to deliver the infant to Semele's sister, Ino, for upbringing. Later, Dionysos would redeem his mortal mother and bring the shade of Semele back from Hades to Olympus to reap the benefits of being a goddess. This power to transform, which is expressed even in his ability to change his own form to that of a lion, horse, or serpent, becomes one of the signatures of Dionysos.

In the Greek pantheon, Dionysos' ways are opposite those of Apollo, who represents ultimate order and precision in the world. (Apollo is also the conservative brother of Hermes.) The philosopher Nietzsche pits the Apollonian and the Dionysian as the instinctive struggle of the Greeks. In *The Birth of Tragedy,* Nietzsche posits Apollo as a god of difference (individuality) and integrity, and Dionysos as the wilder god of unity (togetherness) and intimacy. Dionysos' spirit is one of cooperation and facilitation between society and nature—embracing the edge of chaos—rather than personal aggrandizement and authoritarian rule of order.

As Graham Parkes writes, "The Dionysian drive dissolves the barrier of the individual self in two ways: it breaks down the barriers separating it from other human selves' leading to a feeling of oneness with the social group or—by extension—with the human race; and it also dissolves the boundaries between the individual human and the world of nature, conducing a sense of unity with the cosmos."[8] This seeking of common ground with the cosmos and all its creatures is ultimately beneficial for humankind. The Dionysian way associates itself with intimacy in nature and participation in the universe's unfolding.

By contrast, according to Nietzsche, "It is Apollo who tranquilizes the individual by drawing boundary lines, and who, by enjoining again and again the practice of self-knowledge, reminds him of the holy, universal norms."[9] But these prescriptions for an orderly life and the "proper" organization of society are all too easily appropriated by despotic or fascist rulers and religions (even philosophers have misread Nietzsche as promoting the aggrandizement of power). Moreover, these touted universal norms become the projections of a transcendent realm that find their apotheosis in the philosophy of Plato, especially in his Theory of Forms and the benevolent but authoritarian rule of the Philosopher King in his *Republic,* which was Plato's vision of the ideal society.

The "holy, universal norms," in their rigid application—anathema to anyone who espouses freedom of action and wishes to live in a world inspired by Hermes, Dionysos, Māui, Coyote, and the like—disengaged humans from the natural flow of universal evolution. The creative force and cooperative self-organizing potential of humanity are suppressed as people become captives or slaves within authoritarian social systems. In the controlling Apollonian ethos, still so widely adopted in modern societies—in governments and corporations and aggressive religions—the lessons of Hermes, Dionysos, and their counterparts, once guidelines to human participation with nature and channels of openness to emergence, have faded in our imaginations and have become shadowy tales from our forgotten primitive ancestors. And the fate of societies that fall into authoritarian thrall, erecting Apollonian barriers of difference, is to follow a trajectory toward stasis, the death of innovation and emergence.

Figure 2.1 Hawk and Raven, representing orientations of self-potential from Apollonian versus Dionysian perspectives. Dialogue is a translation of insights by Arthur Schopenhauer in *The World as Will and Representation* (1818). *Credit:* Drawing by Susan Culliney.

However, myths of participation have recurred throughout history. They often take hold for a time in countercultural movements. Indeed, earliest Christianity was a model for cooperative cultural intimacy within a revolutionary religious framework. Ultimately, this hopeful movement that began with its strong, intimate, participatory social ethos was perverted and captured by an authoritarian elite that promulgated self-serving norms as holy and universal. Unlike Jesus, whose mythic powers extended even to cooperative accord with natural forces, the exalted vicars were unable to control turbulent seas of dissent that split the church in sundry directions and largely abandoned the true Christian message in favor of authoritarian commandments. (We further examine the hijacking of social intimacy in the West and Middle East by organized religion in chapter 10.)

In *The Birth of Tragedy*, Nietzsche points out that "Apollo embodies the transcendent genius of the *principium individuationis;* through him alone is it possible to receive redemption. The mystical jubilation of Dionysos, on the other hand, breaks the spell of individuation and opens a path to the maternal womb of being."[10] We suggest this path to the maternal womb of being nurtures and guides the fractal self, as glimpsed by Nietzsche, and the maternal womb of being represents the human potential for intimate engagement in universal evolution. As human beings, we are creatures with natural religions that come from much deeper roots than any "revealed" credo. A primary goal of embracing religion is to become part of something greater than ourselves. From the evolutionary perspective of intimacy as sensed in the Dionysian myth, we aspire to participate in the becoming of the universe.

The Way of the Shaman

Coyote, Māui, Hermes, and Dionysos all represent the need for episodes of creative turbulence as a disruptive force out of stasis toward innovation, or the reshuffling that is necessary to yield opportunities for emergence of something new and progressive. The manifestation of the new is the feedstock of evolution and, often, a prerequisite for the health of any unfolding system and its diverse constituencies. Ancient men and women were aware of this in their myths, rituals, and practices because of their intimate relation with the cosmos; they defined themselves in relation with the natural phenomena of the sky and earth; they found themselves in a continuum of flow of *mana* from the material to the spiritual. Among these people in many cultures, the adept, or shaman, or natural sorcerer (who is a prototype of a fractal self) had the deepest understanding of the connections of human beings to all things. The best of shamans would never seek the security of the province of Apollo, but rather would allow themselves to be drawn into the deep attractor of participation in nature, enjoying the sense of discovery and perhaps anticipating emergence—the province of Dionysos.

Shamans, as the classicist E. R. Dodds defines them, have "received a call to a religious life. As a result of this call [they undergo] a period of rigorous training, which commonly involves solitude and fasting, and may involve a psychological change of sex."[11] Once the shaman emerges from this religious training, he possesses, according to Dodds,

> the power, real or assumed, of passing at will into a state of mental dissociation. In that condition he is not thought . . . to be possessed by an alien spirit; but his own soul is thought to leave the body and travel to distant parts, most often to the spirit world. A shaman . . . has the power of bilocation. From these experiences, narrated by him in extempore song, he derives skill in divination, religious poetry, and magical medicine which makes him socially important. He becomes the repository of a supernormal wisdom.[12]

Thus, shamans seek a balance between the mythical/magical and the real Earth; that is, they are students of the plants, animals, rivers, and the rest of nature. They instinctively feel and see magic in the state of nature and have an intensified intimacy with nature beyond any of their lay counterparts in society—their selves are fractally enmeshed with the patterns of the natural world. Not only do shamans move between the normal and supernormal, between the human and natural worlds, they also develop a heightened state of empathy with their fellow human beings.

The word *shaman* comes to us from the Tungus language of Siberia and means "he or she who knows." Shaman is much like the term *mana* in its appropriation by anthropologists to refer to a wide range of cultural experiences and practices. In his classic work *Shamanism: Archaic Techniques of Ecstasy*, Mircea Eliade, the great historian of religion, concludes that shamanism is at the foundation of all religious and spiritual ways of life. In terms of intimacy, shamanism then can be found to be especially compatible with such religions as Buddhism, Confucianism, Shinto, and Daoism, and once spanned Asia, Oceania, the Americas, Africa, and the Indo-European region and still remains in many of these places. Shamanism even persisted in the Eleusinian mystery cults of the ancient Greeks and later in Roman religion. By definition, shamanism then is intimate and inclusionary, a profoundly widespread cultural attractor. In this sense, shamanism is an aspect of religion and the religious experience based on the mystical integration of the shaman and his or her extra-dimensional surroundings. These extra-dimensional surroundings often included passage to the afterlife. As a guide of souls, or *psychopomp*, the shaman would be responsible for the safe passage from the normal to the supernormal or from the natural to the supernatural. In shamanic lore, according to Eliade,

> healer and psychopomp, the shaman is these because he commands the techniques of ecstasy—that is, because his soul can safely abandon his body and roam at vast distances, can penetrate the underworld and rise to the sky. Through his own ecstatic experience he knows the roads of the extraterrestrial regions. He can go below and above because he has already been there. The danger of losing his way in these forbidden regions is still great; but sanctified by his initiation and furnished with his guardian spirit, a shaman is the only human being able to challenge the danger and venture into a mystical geography.[13]

Our word "ecstasy" is an interesting one, and it can be applied to the shamanic experience. Following its ancient Greek etymology (*ek* + *stasis*), the word ecstasy literally means to stand outside of oneself. This state of "standing outside of oneself" is an expression of a heightened sense of participation where the self has become enhanced by being a part of the process of the world and its development. As scientist, healer, priest, and humanist, a shaman commonly takes on a plethora of roles in his or her society and blurs distinctions among them until they are woven together in fractal patterns. Conducting experiments with plants and animals gives shamans an intimate knowledge that does not accrue to others. Here is implicit intercession with nature, and to be natural with nature— that is, to understand, flow with, and even appropriately use nature's resources— requires training and skill that reaches its highest expression in the most open and adept of society's members. This knowledge is important for the members of the shaman's social group, without which societies would not be able to make progress in living with the Earth. The shaman as healer gives us our first glimpse of a holistic approach to healthy living in participation with nature and in cooperation with the community. The results of shamanic intercession in relation to the sustainability of society are even more important than the shaman's social leadership in the community or even her or his system of healing.

Reaction and Resistance: Temptation toward Authoritarianism

With its emphasis on transcendence, Christianity (as well as Islam, or any religion that suppresses or destroys indigenous religious practices and sensibilities with either relentless conversion or mass murder) eliminated much of shamanic practice with a heavy, often brutal footprint wherever they touched down.* Today, shamanism survives mostly where aboriginal descendants have somehow managed to retain their cultural heritages and ancient practices of intimacy. The transcendent focus abandons the many personified gods and goddesses and demi-deities of the shamanistic world that are devoted to specific phenomena and processes (such as lightning and thunder). The transcendent perspective gives us a supreme being, existing virtually beyond our imaginations. The dominion of the transcendental God is characteristically one of perfection—everything the human is not—and this God's powers become unapproachable to humans, for He is omniscient, omnipotent, omnipresent, and so forth. Since God is now associated with all that is unearthly, He resides in the sky, or even beyond the sky, in a heavenly realm of perfection and bliss. This God typically becomes associated almost exclusively with the masculine, and His Word or scripture may dictate the suppression of the feminine. But His supernatural status is starkly segregated from that of all human beings, now considered to be mired in nature, who seek salvation from beyond for their own transcendence and, in some cases, for their own personal glory.

Along with this consolidation of the many gods to the One comes the transformation of the idea of the intimate self to the idea of the unified self of integrity. This development, which occurs primarily but not exclusively in the West,[14] would be promoted in later Western philosophy and religion as an advancement, improvement, and progress over the more primitive and less aggressively promoted perceptions and beliefs of how to negotiate life and death with the planet.

To probe the origin of transcendence and its seductive power of authoritarianism and influence on Greek philosophers and the Abrahamic traditions, we might reflect on the story of Father Sky and Mother Earth in Hesiod's *Theogony.* Hesiod's story is one of the first creation myths of the West: how Sky, the male principle, came to seek his revenge against Mother Earth. In the backstory of the Greek pantheon, long before the advent of the familiar names—Zeus, Apollo, Hermes, and the rest—Hesiod narrates the following cosmogony: "First of all, the Void [Chaos] came into being, next broad-bosomed Earth, the solid and eternal home of all, and Eros [Desire], the most beautiful of the immortal gods, who in every man and god softens the sinews and overpowers the prudent purpose of the mind."[15]

---

*We use the word "transcendence" here to indicate the philosophical and religious tendency to emphasize the otherworldly, or the tendency that pulls away from intimacy toward integrity, that is, the tendency that pulls us away from conceiving a self that is defined in connection to natural phenomena and evolutionary processes.

Echoing the power of *mana*, the early Greeks gave Eros (who instills desire in the form of a yearning for unity) a universality that was coterminous and contemporaneous with the origin of the universe itself. In a primordial-mythical way, Hesiod is giving ontological status to the power of Eros in the sense of the cooperative constant, for sexual reproduction is a fundamental example of the participatory principle operating not merely in humans but in nearly every complex organism on Earth (see chapter 5). As the first of the ancient Greeks to promote the primacy of Eros, Hesiod recognized the emergence inherent in the self joining forces with the other, and he extended the creative power of cooperative unity across the spectrum of nature—life, Earth, cosmos—and into the realm of the gods and goddesses themselves.

The communion of Mother Earth, also known as Gaia, in her role of physical creation, and Eros, as the creative principle of intimacy, represents the necessary condition for embryonic expansion in the void of Chaos and all future emergence in the universe. This is where the world becomes self-organized through the interaction of male and female. Although Eros' reproductive potency pervades all relationships, it is diffuse in nature, and earthy Gaia has a mind of her own. Out of unanswered passion she creates a consort who will remain close and visit her intimately each night. He is Ouranos, also known as Sky. Thereafter she will lie beneath Ouranos, who will become the father of her children; but their union is destined to lead to troubling turbulence and eventually a profound shift to a transcendental worldview in the concept of a powerful, vengeful deity whose realm is the exalted, isolating sky.

The break begins with the perversity of Ouranos. He is a reluctant father: "As each of his children is about to be born, Sky would not let them reach the light of day; instead he hid them all away in the bowels of Mother Earth. Sky took pleasure in doing this evil thing."[16] The refusal of Sky to assume parental responsibility evokes the classic alienation of authoritarian father figures in Western culture—a resistance to foster the natural tendency for emergence in children. His retreat into the stance of overbearing control and cruelty will challenge Gaia to the maternal defense of her offspring—family discord emerging from parental polarization. Father Sky's pretensions to omnipotent patriarchal status make him the prototype of Jehovah and Allah (God, the Father), although those later godheads, perhaps out of self-preservation, have not allowed themselves to become intimate with Gaia.

Ouranos' downfall is precipitated by the youngest and boldest of Gaia's children—and one of the first trickster gods—whose name is Cronus (he will later become the father of Zeus). Gaia manages to give birth to Cronus, who turns out to be a highly precocious lad. Out of her own raw materials and divine metallurgical skills, Gaia fashions a gigantic sickle and places it in Cronus' hands. She conceals Cronus in ambush, as "Huge Sky came drawing night behind him and desiring to make love, . . . he lay on top of Earth stretched all over her. Then from his ambush his son reaches out with his left hand and with his right took the huge sickle with its long jagged teeth and quickly sheared the organs from his own father and threw them away, backward over his shoulder"[17] into the sea.

The Triumph of Transcendence

After Cronus sheared the descending genitals of his aroused father and tossed them over his shoulder into the sea, the Aegean accepted them and took the organs into her waters. In spite of Ouranos' attempts to dominate Gaia and subjugate her according to his own selfish desires and wishes, he is surprised she fights back, and perhaps his disbelief is intensified by the recruitment of his own son to act so pugnaciously against him in such a premeditated way. Gaia's movement forward is inexorable, and giving birth is a principle of her nature that exalted Sky somehow failed to understand; the urgency of her own needs, most immediately her labor leading to birth, are inevitable and universal. The labor pains of Gaia to bring her children forth can be seen to embody creative turbulence to the point of emergence. This sense of emergence is anticipated in Hesiod's story.

Hesiod continues, "The drops of blood that spurted from them were all taken in by Mother Earth," and "as for the organs themselves, for a long time they drifted around the sea just as they were when Cronus cut them off with the steel edge and threw them from the land into the waves of the ocean," but "then white foam issued from the divine flesh, and in the foam a girl began to grow." This "tender and beautiful girl" emerged a goddess, and "round her feet the green grass shot up. She is called Aphrodite by gods and men."[18] Hesiod seems to have been aware of Mother Earth's amazing and natural ability to transform her world from violence to love, to ease into the transition from discord into harmony, and to evolve from disunity into unity through her innate power of guiding emergence and creating capacity for self-organization. Aphrodite, the goddess of love, comes into the world through an extreme turbulent act of castration that is creatively transformed into a beneficial event of evolution. "Great Father Sky called his children the Titans, because of his feud with them: he said that they blindly had *tightened* the noose and had done a savage thing for which they would have to pay in time to come."[19] We suggest that Sky's revenge led later to the concept of a vengeful male deity residing in a remote heavenly realm, whose anger will be directed against the Earth and her inhabitants. Thus the invention of transcendence can be seen as a religious manifestation of Sky's castration.

The pronouncement of Sky's revenge on the Earth and her children can be understood metaphorically, even mythically, as a premonition of the subsequent appeal of transcendence in the West. After sensible (and prescient) attempts by Presocratic philosophers to understand nature (*phusis*), Plato took up Sky's revenge and rendered the Great Beyond (his Intelligible Realm) as being vastly superior to the Sensible (what we would term the tangible universe), where he relegated the inquiry of most of his predecessors. Plato's influence on early Christianity was significant and provided Christianity's original and primary transcendent orientation toward redemption, agapic love, and the association of Jesus as God. All this would lead Nietzsche to later proclaim that "Christianity is Platonism for 'the people'."[20] This orientation would also be triggered by the work of Plato's brilliant student Aristotle, who came to exert a profound influence on later Christianity and Islam. The Neoplatonist philosopher Plotinus, who believed in the absolute transcendence of the One (God), even proceeded to a more extreme view by speculating that Plato's Intelligible Realm was actually beneath the One and overflowed from it—thereby suggesting an even more remote supernatural realm.

From Plato, the West inherited a vision of absolutes, with the Form of the Good held as the supreme universal form that is attended by the Forms of Equality and Justice. In his *Timaeus*, Plato gave us a creator in the guise of a Demiurge, a craftsman that "works" for the *demos*, the people, by bringing form to the formless (special creation). Aristotle then speculated on an Unmoved Mover, or first cause of all there is, because the idea of a series of causes infinitely regressing back in time was unreasonable and inconceivable to him. And all of this despite his ancestors who conceived of atoms! (See also chapter 1.) And as these monotheistic prototypes came forth, corresponding conceptions of an immortal soul arose for the first time. This soul was to be found somewhere and somehow beyond the body and its phenomenological world of flux and change, of evolution. This soul was intimately linked to the eternal, the Intelligible Realm for Plato. This soul was inferred to be tripartite in nature, with its rational part held as superior; and it is only that part of us that ultimately can know the universal Forms. Following his teacher's lead, but tweaking Plato's conception, Aristotle's idea of the soul consisted of three powers or capabilities, rendering it even more individuated, unified as a self, and differentiated as a separate entity than Plato's. This sense of soul became the foundation for the integrity self that supplanted the version of the intimacy self we find in Homer.

Alfred North Whitehead once proclaimed that "the safest general characterization of the European philosophical tradition is that it consists of a series of footnotes to Plato,"[21] and we can begin to sense that this is only a modest overstatement, for there is so much truth in it. Whitehead is only echoing Plutarch's earlier decree that "Plato is philosophy, and philosophy is Plato." The influence of Plato, with his long line of converts, followers, and interpreters, has been a persistent and ongoing philosophical foundation of the West's religious worldview. This trajectory toward transcendence widely replaced the natural wisdom of primordial myth and its evolutionary outlook—the habitat of free-thinking tricksters and shamans and earliest scientists—with a far more superstitious myth of revelation that stranded much of humanity on a sterile shore in the doldrums of a received and absolute Truth. But intimacy-minded thinking did not decline everywhere. Perhaps with direct lines of inheritance from ancient mythological memes far to the east of the Mediterranean world, very different philosophies arose that managed to avoid the transcendental trap. Societies that long remained apart from Western influence maintained connections with "the divine" in nature and in its continuum with human nature. Their bodies of philosophical thought conceived an understanding of the self-organizing principle and the fractal self with clarity and depth and richness of metaphor, and we will return in detail to their approaches to these ideas in concluding chapters. Moreover, those non-Western philosophies seem most resonant with current scientific models of the complexity-building cooperative pathways of organic and biotic evolution, which we outline in part 2.

PART II

The Ascendance of Cooperation

The Quickening of Chemistry

Life, this anti-entropy, ceaselessly reloaded with energy is a climbing
force, toward order amidst chaos, toward light among the darkness of
the indefinite, toward the mystic dream of love between the fire which
devours itself and the silence of the cold.

—Albert Claude, Nobel lecture

Views of the evolving universe—both scientific and humanistic—are richly imbued with the idea of cooperation. We suggest that through most of the history of our universe, cooperation has held a slight edge over that other much-ballyhooed Darwinian ruling principle in nature, competition. This hypothetical edge is a condition we have called the cooperative constant, and we further maintain that without this critical imbalance, the phenomena of complexity and emergence—the sources of all nature beyond subatomic particles—would not exist.

In this chapter we pick up the scientific story of evolution. Recall that in the earliest moments of our universe, as cosmologists have it, a colossal competition between matter and antimatter was settled in a kind of secondary Big Bang. Cooperation was impossible at that juncture, so perhaps competition is primal—or at least the antagonistic identities of self and other were primal—and at the beginning of everything this distinction was inevitable. Moreover, in this earliest competition, the universe was made simpler, not more complex. But once the issue resolved, affinitive entities in the universe were free to attract and assemble, associate, facilitate, and cooperate, rise above the leveling action of competition, and generate emergence on progressively higher levels: chemical, biological, and social.

Now we return to cooperation and examine its constructive power in what might be termed *ascendant chemistry*—the self-organization of molecules that led through pathways of emergent complexity to the threshold of biology and the evolution of life on earth. Indeed life arose as an assemblage of complex molecules with strong cooperative tendencies within and among themselves. As we noted briefly at the end of chapter 1, early cosmic chemical complexity (the forging of elements in stars and the onset of chemical reactions) took an enormous leap forward with the propensity of carbon atoms to bond with one another (together with atoms of many other elements) and to build endlessly variable

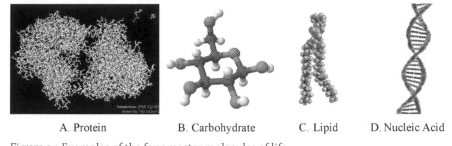

A. Protein	B. Carbohydrate	C. Lipid	D. Nucleic Acid

Figure 3.1 Examples of the four master molecules of life.

Credits: A, MIT.edu, author Tom Vickers, public domain. *B,* Wikipedia, Wikimedia Commons, author Ben Mills, public domain. *C,* publicdomainpictures.net. *D,* slideshare.com.

molecular structures—most notably of the "master molecules," known as the four basic complex chemicals of life: proteins, lipids, carbohydrates, and nucleic acids (DNA and RNA).

These very materials, however, do not suddenly materialize as from the brow of Zeus. They themselves are intricate composites. Between the simple, covalent bonding tendency of carbon atoms and the synthesis of a macromolecule such as a protein is a stunning, shape-shifting process of polymer construction that seems evocative of a cooperative constant operating at the chemical level. During the assembly of a polymer,* whole smaller molecules (monomers, or building blocks) are induced, in particular aqueous environments of the proper mixture of salts, a certain range of acidity, and so on, to form bonded associations that ultimately relax into stable, composite configurations. According to one prominent hypothesis, nano-scale patterns of electrical charges, such as occur on tiny flakes of clay in warm waters, may have provided the first catalytic attractions, bringing molecular building blocks together in favorable positions for linkages to form.[1] Then, in myriad combinations and permutations within the four basic types, the macromolecules that participate in life began to interact with their chemical surroundings, build composites, and generate dynamic functions and cooperative behaviors among themselves in amazingly emergent ways.

This is a surprising trend in our universe: that self-organization of matter and energy builds complexity in surrounding seas of disorder. The disorderly mandate is dictated by a basic tenet of thermodynamics—namely that dissipation and randomness must forever increase after the Big Bang. Of course, our small

---

* Technical note: Many lipids, like proteins, complex carbohydrates, and nucleic acids, are large composite molecules but not true polymers, which require assembly in the form of chains of many small molecular building blocks, or monomers. However, lipids do commonly associate, forming vital and dynamic structure, such as the ultrathin, intricate membranes that surround cells and various parts of cells.

corner of the cosmos, evolving to an organic epiphany, requires vast inputs of raw energy from the sun in order to generate ascendant outputs of order and produce entities that gain influence in nature—the fruits of progressive emergence.

Throughout molecular evolution, the millions of varied instances of chemical cooperation that arose spontaneously were tested by natural selection, initially simply for their stability or survivability, and survivors provided feedstocks for the raveling of greater complexity—from atomic bonding to reactions of monomers amid water and ionic reactants to the sophisticated catalysis promoted by protean macromolecules. If survival of the fittest* (a phrase not invented by Darwin) is a tautology, as identified by many logicians, it is nonetheless logical to infer that through the eons and within the Earth's pre-biotic "oceanic soup," Darwin's great principle sieved and scythed its way through ever-more-potent raw material to favor the fast-acting, strong, flexible, adaptable, and, especially, the cooperative among macromolecules. Forged through bouts of competition, hierarchies of cooperation inexorably emerged and became robust.

But the innovative side of evolution in its prebiotic phase—its ceaselessly creative potential—seemed always serendipitous, highly dependent on interactions of the various master molecules. Aside from the simplest laws of chemical bonds between atoms, there was no predicting or directing the appearance of some new form of composite association, catalytic action, mutualism—the inevitable emergence that would take a progressive, local, chemical mixture to the next level in some nurturing nook on this not-too-turbulent planet. Yet, once closely knit molecular systems achieved the status of life, the universe itself became an egg and began in the crudest sense to anticipate wholly novel properties—from the organic, self-regulating computation of living cells to mind that has begun to span the cosmos.

Life Finds Its Way

One of the seminal realizations of late twentieth-century biology was that life did not arise by chance. Until quite recently, biologists were still struggling to explain what many scientists viewed as a vast improbability: the generation of amazingly complex, self-replicating systems of giant molecules that cooperate in the forms and functions of living cells. That the structural intricacy and synergistic properties of those molecules seem to have come together *randomly* in the ultimate chemical emergence of life is plausibly viewed as impossible. To extend one popular myth, not only is the spontaneous appearance of a composite and extremely complex system such as a living cell roughly equivalent to a Boeing 747 being assembled by a tornado traversing a junkyard, but this aircraft also features an onboard computer, with software that nobody programmed, and is

---

*"Survival of the fittest" was first used by the British scholar Herbert Spencer in his little-known work, *Principles of Biology*, in 1864, after he had read Darwin's *On the Origin of Species*. Darwin adopted the phrase in a later edition of *Origin*.

then seen to fuel itself, start its engines, and take off without a pilot.[2] Creationists love to cite this as a parable of ridicule against biotic evolution. In his 1998 book, *Darwin's Black Box*[3] (touted ever since by creationists and "intelligent design" advocates), biochemist Michael Behe first posited what he termed the irreducible complexity of certain structures in living cells. Behe's point was that some of the molecular architecture and economy of a cell appears too complicated to have evolved. (We will return more specifically to Behe's argument later in this chapter.)

The false premise pushed by creationists and accepted all too readily by many intelligent but poorly informed people with a general interest in the subject is that each step in an evolutionary process proceeds out of randomness or equal opportunity. The illogical idea that inputs, or initiating actions, must always, at every step of the way, happen by pure chance, and that *outputs*—that is, products or consequences—are all equivalent in fitness or influence, has nothing to do with the way nature operates in our universe. Tremendous insights of twentieth-century science, and especially biology—in molecular and cytological research, genetics, embryology, and ecology—have shown that nature is intensely hierarchical, with functional novelty *progressively emerging* from the cooperation of systems. Adaptive value, efficacy, and utility (in the sense of Darwinian fitness) then are favored as a result of natural selection—all of this in an irregular, stepwise (but not always gradual, see chapter 4) buildup of form and function. New organic capabilities may evolve from sometimes serendipitous hand-me-downs (in the case of structures that begin by serving one function and later, with descendant modification, become co-opted for another),[4] but the assemblies of systems that led to living forms and their working parts were never random and trended strongly toward diversity, often toward complexity.

Here we should again reflect that without cooperation at any level, gains in complexity are few and far between. Haphazard chemical reactions between colliding molecules or random mutations in living cells are rarely constructive. Left to themselves, random processes are commonly neutral or may be destructive. Competition in nature forms patterns of various kinds through antagonistic spatial arrangements of the competing entities; often such outcomes generate separation and sometimes simplicity. While it is true that, as Darwin noted, the "struggle for existence" may temporarily increase fitness in the survivors of contests between closely matched competitors, the outcome may restrict future evolutionary diversity. Competition may enhance and specialize certain adaptations (of the "fittest" competitors), but, if one-sided, it narrows the field of options, leading toward dominance of one life-form over others or extinction of the less fit in the context of some local environment. In fact, ecology (and more recently paleoecology) has provided evidence that it is the *avoidance* of competition that has the effect of increasing complexity in the form of biodiversity. The avoiders are individuals already progressing (passively, through random mutation tested by nonrandom selection) toward developing an adaptive shift in their biological nature, fitting them for an adaptive landscape.[5] This is a major mechanism in the microevolution of organisms, leading to speciation itself, and beyond, in the emergent richness of ecosystems.

Molecular Mutualism

Among macromolecules, it is in the nature of many to assemble cooperatively. Proteins and nucleic acids (DNA and RNA) especially exhibit this far-reaching property, among themselves and in combinations, for example, in the cell organelles called ribosomes.[6] Fractally congruent shapes and the emergent properties of catalysis and "irritability" in proteins are well-known, and such "smart" proteins clearly organize and arrange themselves spontaneously in nature. God did not have to fuss with them; alien super-scientists did not whip them up in an intergalactic lab. Cooperative assembly of complementary molecular shapes, guided by mutual attractions of minute electrical fields (again distributed wholly naturally through the chemical building blocks—amino acids, in the case of proteins), generates iteration after iteration of composite structure. At each new level, antecedent structures can contribute to the next generation. Any useful or self-perpetuating function in the context of the structure's immediate surroundings will be favored by selection. Eventually, such molecules find their way into efficacious combinations, spatial and energetic relationships, perhaps following similar basic principles of interactive development as organizes species in an ecosystem such as a coral reef, a grassland, or a forest.*

There is a clear sense, albeit not overtly recognized, in which Buddha nature—the state of connectivity of all things—extends with ease into chemistry. Most Buddhist scholars trace the path of enlightenment in the world as passing through nested sets of connected principles. As if in a series of fitted Chinese boxes, all natural systems are linked: humanity is connected to animate nature and followed by vegetative manifestations eventually converging, in the classic view, to a relationship with stones. Stones have traditionally been at the bottom of this worldview. Nevertheless, even stones on a landscape such as a riverbed or a garden are seen to have a form of wisdom as they channel flows of water or induce growth patterns of plants' roots; they may have a backstory, perhaps tracing to their origin and ancient spatial configuration when they were embedded in mountain strata, and the study of Buddha nature attempts to read their narrative and bridge it to others'.[7]

How much more dynamically connected to animate nature are substances like proteins! Buddhist biochemists might suggest to geologists that proteins have many more and varied stories than stones. And if, indeed, stones record epic tales of the passages of the earth, proteins' stories are almost as timeless, but exquisitely compressed, like haiku, extraordinarily intimate and astronomically diverse. Theirs are stories that come alive in myriad patterns. Nucleic acids, too,

---

*Of course molecular assemblages are subject to electromagnetic and sometimes quantum effects that are not directly involved in ecosystems.

have their narratives, simpler and perhaps more ancient than those of proteins.[8] As we shall see, some of RNA's oldest stories, recently uncovered, are characterized by intricate twists.

Even in the formative biochemical world of the early Earth, various molecules connecting to others and then preserving the connection resulted in the storage of information. The development of a simple series of chemical reactions leading to a product whose presence then stimulates, or catalyzes, the repetition of that same series is known as an autocatalytic set.[9] Thus nature remembers, even at this most basic level. It does not have to start over at every new juncture. This molecular record-keeping and tracking of progress is the very simplest process of life. It starts before there is life, then is co-opted by living cells. The result is a storing up of some substance or level of energy, a chemical tension and potential. And such a self-stimulating, self-preserving quality represents a significant repository of information in a tiny locus of the universe. This is directly on the pathway toward life.

In the beginning of organic chemical stirrings in the "oceanic soup," such a simple positive feedback process may have acted as a tremendous multiplier of certain chain reactions. The increase of self-aggrandizing products may have been limited only by hostile physical conditions or events—sharp gradients in water temperature or acidity, for instance, or limitation by exposure to radiation, including visible light. But there would have followed a gradual change to more sophisticated regulatory processes of protobiotic pathways of synthesis— chemical-on-chemical regulation by shape-shifting proteins—that prefigured the emergence of the controlled, comprehensive, if not always precise, homeostasis that is a defining property of living cells and organisms.[10]

Thus, formative and primitive chemical feedback mechanisms—those that accelerated the buildup of specific substances by autocatalysis—were the first modes of organic reproduction and would have triggered classic Darwinian competition. Selection would have favored the superior replicators, eliminating others or relegating them to marginal places of less-than-ideal physical conditions. In this, the balance again would have tipped toward simplicity; raw aggrandizing power was a chemical process of chain reactions on the threshold of life. Competitive chemistry in the oceanic soup may have often retreated from the potential of complexity, especially of emergence that springs from the openness of a world with numerous options. But cooperation showed its staying power in the long run. The metabolism of simple cells would come to harbor regulating versions of both proteins and RNA galore, all ultimately tuned to adjusting the rates of production of key substances, maintaining balances, and modulating flows of reactants—a complex chemical intimacy that translates into smooth functioning, ultimately, of energy processing, growth, and reproduction of cells themselves. The profound emergence of cellular life would have been impossible in a world without the cooperative constant operating at the *molecular* level.

In the case of proteins, the vital catalytic activities of enzymes, with their regulation of cellular metabolism, are nowhere near the limit for emergent biochemical chutzpah. Several kinds of so-called motor proteins have affinities for much smaller high-energy molecules. In living cells these proteins, among numerous others, show a great preference for a little chemical sparkplug called ATP (adenosine triphosphate) that can contribute high-energy electrons to destabilize many chemical substances and thus drive chemical as well as physical reactions. ATP binds to a key location on a motor protein's surface and releases a tiny jolt of energy. The protein reacts with a sudden change of shape; it gives a mechanical twitch or push or pull against some adjoining structure. When numerous such proteins are linked in orderly, cooperative arrays with others, the twitch produces a stunning emergent summation—a *Paramecium* swimming, or a greyhound running.

The Intimacy of DNA and RNA

Nucleic acids also function in cooperative structural and functional arrangements involving active attraction, synergistic information storage, and communicative output between self and other. Most famous in this context is the complementarity of the strands of the DNA double helix itself, discovered by Watson and Crick in 1953. Following on the heels of that discovery of the simplest property of DNA came further research by numerous individuals that led to our understanding of the code by which the polymeric sequence of DNA's building blocks (segmental nucleotides) translates into the sequence of amino acids in specific proteins.

Out of this work emerged the central dogma (as it has been known ever since) of the molecular basis of heredity, which crystallized in a gross simplification called the one gene–one enzyme theory. In this theory, the genetic code is passed (transcribed) from DNA to the code-carrying messenger RNA, whose nucleotide sequence (in triplets) is translated in linear register to the sequence of amino acids in a protein chain. This dogmatic view of biology pervaded the minds of top scientists in the field: that the essential information for the construction and operation of cells and organisms simply passed from DNA, the "master molecule," downward to RNA, and finally to proteins. To James Watson and some others, molecular biology had uncovered the First Commandment; all else in biology could now be reduced to the irreducible gene. For a time, DNA came to be regarded as a kind of godlike authoritarian substance that invariably dictated the code, and each gene segment along the DNA translated into a particular protein that had a fixed task, somewhat like an Orwellian vision of the cell.

Before the 1980s, however, research discoveries had begun to show that the central dogma was a greatly oversimplified view of a much more intimately ordered complex system, in which proteins and RNA provide input to the transcription, translation, and application of the genetic code, altering the products of genes in myriad variations. The new views of macromolecular communities in cells with numerous pathways of information flow, and with DNA no longer in its unalterable, dictatorial mode, have been scientifically confirmed, with ever more examples accruing for over three decades.

If the ancient Greeks had discovered DNA and RNA, they would have linked those molecules to Hermes, the propagator of language and demigod healer, as the obvious patron of the polymers that encode—in caduceus coils—biological form and function and carry vital messages that ultimately manifest in health and fitness, metabolism, growth, and life itself. In the spirit of Hermes, nucleic acids have continued to reveal surprises. RNA, the deceptively single-stranded siblings of DNA, became known for several versatile supporting roles in the conversion of life's DNA blueprint to protein structure and action. Three classic functions* had come to light by the 1960s.

However, in the early 1980s, a stunning further role of RNA surfaced—a capability that suggested RNA's primacy in approaching the threshold of life itself in the primordial dreamtime of the biosphere. This was the discovery of RNA catalysts, now called ribozymes, which led to a conceptualization of life's debut in an "RNA world."[11] If any molecule deserves to be called a bootstrap to life, RNA perhaps best fits the description. And, incidentally, ATP, which has been called the universal energy currency of life—the sparkplug substance from the earliest eon of biological chemistry and vital to maintaining virtually all life on Earth—is little more than a monomer of RNA carrying two extra linked phosphates that stretch the limits of unstable chemical energy in their bonds.

Elementary biology texts often characterize RNA as a single-stranded nucleic acid; this quality is put forth as a key difference from the doubled structure of complementary, helically entwined strands of DNA. However, shapes of RNA molecules commonly depart from a simple linear chain of nucleotides resembling a line of conga dancers. Just as in a strand of DNA, an RNA molecule consists of four versions of nucleotides—the monomers that bond in a chain to form the polymer. One of the RNA monomers differs in its identifying base component from one of its DNA counterparts. Otherwise RNA starts out looking virtually the same as a single strand of DNA. But it doesn't stay that way for long. Whereas two DNA strands normally engage one another in their twining embrace, RNA practices a kind of significant-other-seeking behavior among nucleotides along its own single polymeric chain. In a complementarity that can never be fully consummated, RNA loops back on itself. The bases of RNA, like those of DNA, are amenable to pair-bonding, and the rules are the same: only two specific pairs of the four nucleotide bases are capable of coupling.

Now the variable sequencing of the four nucleotides of any significantly long RNA strand comes into play. The stringy molecule can writhe and twist and loop around itself, and anywhere that the complementary base pairs along the chain can approach each other within a short critical distance, they can connect with the same types of chemical linkages that characterize the cross-bonding of two-

---

*Three classic cellular roles are observed for RNA: *ribosomal RNA* structurally shapes and catalyzes some functions of ribosomes; *messenger RNA* transcribes (copies) the genetic code of DNA into an RNA sequence of nucleotides; *transfer RNA* binds to amino acids and delivers them to precise catalytic sites on messenger RNA as its genetic "message" is being "translated" by a ribosome. See any biology text.

Figure 3.2 A folded RNA molecule with catalytic activity. Hammerhead ribozyme. *Credit:* Wikipedia-Public Domain photo, https://upload.wikimedia.org/wikipedia/commons/2/28 /Full_length_hammerhead_ribozyme.png.

stranded DNA. Thus RNA is capable of forming multitudinous varieties of complex, looping shapes out of its partial complementarity within a particular chain of nucleotide sequences.

The resulting shapes often resemble weird hairpins, bent paper clips, outlandish wire sculptures, and three-dimensional keys to strange futuristic locks. Where an RNA strand folds back on itself to form a doubled section of more than a few nucleotides in cross-bridged linkages, that section of the molecule typically goes into a helical twist like that of DNA. Thus, complementary and noncomplementary parts of the strand converge and diverge like the joinings and separations of braided streams, only in three-dimensional space.

It was those many varying shapes of RNA molecules that led researchers to speculate that, like enzyme proteins, RNA could harbor catalytic potential, momentarily attracting and holding various smaller molecules with shapes that fitted into a particular RNA's pockets of loops and twists, thereby inducing reactive configurations or tensions for chemical change within those smaller molecules. Because RNA, with its four nucleotides, cannot achieve the truly astronomical diversity of shapes and intricacies of molecular-scale electrical fields found in enzyme proteins, whose primary chains can have up to twenty amino acids in play, the realm of RNA catalytic activity is smaller. Nevertheless, out of this class of RNA molecules, the ribozymes are significant players in the emergent chemistry of living cells.

One of the more recently uncovered RNA roles is that of an enzyme guide to locate and enable the alteration of DNA at pinpoint locations, resulting in precision genetic surgery. The biochemical system, called CRISPR, involves a partnership between an RNA guide molecule and a particular enzyme scalpel, and it evolved in bacteria as a defense against viral infection. The acronym refers to the RNA guide: clustered regularly-interspersed short palindromic repeats. The complex term refers to sequences of twenty-four to forty-eight nucleotides that occur in repetitive groups along a bacterial chromosome. As DNA, they are the CRISPR genes, whose products are RNA molecules, not proteins. These repeating sequences had been found in the mid-1980s and characterized in 1993, but their function and significance did not become clear until after 2010.[12]

In their linear array, the CRISPR-coding replicates in each group are interspersed with separate, short "spacers," which have been identified as very small pieces of viral DNA. They are far too short to have any effect on the host, but are mere vestiges left after the molecular scalpel accompanied an activated CRISPR (which "recognizes" the beginning of the viral DNA) to bind and cut the pathogen's genes out of the host's chromosome. Each spacer tells a story of a former viral infection that was successfully terminated.

So, in nature, an attacking virus first merges its DNA into its host's chromosome. Then, before the virus is able to use the host's cellular processes to replicate its alien genes and make the viral proteins that self-assemble into its next generation, CRISPR, if it acts quickly, nips the viral bud. Through their evolution of the CRISPR system, many bacterial species and also archaea (see chapter 4) acquired a type of immune system.

Now molecular biologists can make in their labs thousands of variations on the CRISPR RNA sequence, with tailored complementarity to nearly endless variations of counterpart DNA sequences of any organism on Earth. Injection into target cells of these engineered CRISPRs, having paired them with a DNA-cutting enzyme, makes for extremely accurate genetic surgery at up to many chromosomal loci simultaneously. High speed, unprecedented ease, and much lower cost than ever before has already been demonstrated in manipulating genomes of a widening variety of plants and animals, including human embryos. It should not escape a reader's notice that this presages both extraordinarily promising as well as risky outlooks for the future of humanity (see also chapter 11).[13]

The deceptive simplicity and mercurial versatility of RNA suggest it may well have preceded proteins in leading the way toward the quickening of chemistry on Earth at the threshold of life. Basic components of RNA, DNA, and other biopolymers have now been located in extraterrestrial sources, suggesting these chemical precursors are widespread in the universe and produced by universal processes powered by modest radiant energy near stars. Carbonaceous meteorites contain chemical bases of the types that form the cross-bonded linkages between DNA strands and the folds of RNA. Likewise ribose sugar derivatives are present in samples of such grounded space debris, as reported by NASA chemists and other scientists.[14] Some have even suggested that the Earth was at least partially seeded from outer space with such vital substances of primordial biochemistry in the first five hundred million years after planetary accretion. During this time, nascent planets and moons were bombarded by myriad chunks of matter from space, as the initial huge disk of material surrounding the young sun progressively organized itself into our solar system of ordered planets in settled orbits.

Regardless of the place(s) of origin of the various molecular Lego blocks that built the master polymers, life on Earth likely arose in an RNA world. Among the further capabilities of RNA is promoting its *self-replication*, the simplest form of autocatalysis, as a particular RNA ribbon may act as its own ribozyme. Unlike DNA, whose replication takes numerous convoluted steps that rely on many different enzyme proteins, a self-replicating RNA could churn out complementary copies of itself in a single smooth operation. However, all RNA variations (often sensitively depending on nucleotide sequences) are not equal. Most cannot do the self-replicating trick. Those capable of it have key sequences that induce the ribbon to fold in a unique way, and these reproducing paragons among RNA varieties would have been the opportunistic seeds of the RNA world. Recent research breakthroughs have even demonstrated cooperative activity of RNA, in the absence of proteins or other biochemical materials, in which two different RNA molecules each act as template enzymes for synthesizing one another.[15] This process has shown itself to amplify exponentially and indefinitely, with a doubling time of an hour or less. Moreover, in these experiments, spontaneous variations in the molecules sometimes appeared after many generations, and some of those went on to dominate populations of descendant RNA. As Gerald Joyce, one of the authors of these studies, wrote in 2009, "This is the first example, outside of biology, of evolutionary adaptation in a molecular genetic system." (See note 15.)

It is not unreasonable to suppose that in the ancient RNA world, various other modifications, molecular mutants that lost potency as self replicators, might nevertheless have gained ability to catalyze linkages of amino acid chains, just as do key RNAs in ribosomes of all living cells today. Perhaps these protein-assembling RNAs are functioning fossils of those primeval varieties.

Tiny Bubbles

How did such capable molecules—the genetic information replicators, the energy storers, the energy transformers, and their cooperating others in autocatalytic sets and roughly regulative chemical systems ultimately secure themselves from dissolution and dissipation in their aqueous milieu? How did they break through to achieve the status of living cells? Clues to this stage in life's emergence have come from scientists, such as David Deamer,[16] interested in the behavior of lipids that are naturally occurring in most bodies of water on Earth, from freshwaters to the sea, and are especially concentrated at the very surface of the ocean—the air-water interface. In their considerable variety, including some hybrid forms combined with proteins, marine lipids are the major ingredient of sea foam continuously left by waves on shorelines. Many of these molecules are of the same general type that forms the soap-bubble-thin membranes that enclose all living cells.

Cell-membrane lipids and various sea-foam lipids have a similar structure. They are fishlike in shape, with very long twin tail streamers. The "head" of this sort of lipid typically consists of a phosphate chemical group containing phosphorus and oxygen atoms and is hydrophilic, highly soluble in water. The "tails" of such a phospholipid are hydrophobic; they repel and are repelled by water, causing the molecules to "school" densely together in watery surroundings.

Drifting at the surface of the ocean, or even on a wet beach or wave-washed rocks beside the sea, phospholipids arrange themselves with their hydrophilic heads facedown into water while the tails stick out into the air. Like molecular turf, whole fields or films of phospholipids form in this way. If pushed underwater, as by a breaking wave, many of these contiguous molecules spontaneously find themselves as coatings on tiny air bubbles that soon return to the surface and burst, momentarily leaving a little clot of concentrated lipid, rumpled or folded by its turbulent passage. Reasonable estimates indicate that millions to billions of these bubbles in the size range of biological cells are forming and bursting every second on each square meter of Earth's oceans.

The ultrathin films of lipid on the ocean's surface may even have provided surfaces that brought together prebiotic polymers in the manner that today's cell membrane lipids bind to arrays of protein and carbohydrate molecules and their hybrids. Floating on Archaean oceans, lipid sheets and clots might have concentrated macromolecular chains and hosted autocatalytic sets in an almost homeostatic milieu. Over the planetary sea surface, rafts of "hopeful" biochemical materials could have been continuously agitated and shuffled in endless combinations through eons of time. The whole air-sea interface might have served as an enormous catalytic sheet. And there between the sky and the sea, some of those complex combinations, molecular reefs, might have entrained the emergence of life, held as if in a virtually limitless laboratory shaker or incubator under the misty sunshine and reducing chemistry* of the early atmosphere (which lacked free oxygen), amid occasional discharges of primordial lightning.[17]

One other natural property of the phospholipid films on water is that if they are driven below the surface and fold over themselves without trapping air, they tend to form doubled layers. Now the water-binding heads are immersed on both sides of the sheet, with the tails packed together within the ultrathin filmy envelope known as a bilayer. And when these layered films, themselves, flex and curve amid turbulence, they may produce bubbles that enclose seawater inside instead of air; then they do not automatically rise back to the surface to burst. Deamer and colleagues have pointed to this analog of precisely the lipid structure of the enclosing cell membrane found in every kind of life-form on Earth. Such arrays, called liposomes, would now be stable fully submerged in the water. And in the earliest oceans, over a few hundreds of millions of years, the tiny closed bilayer bubbles provided near-infinite opportunities to harbor the critical range of organic chemical systems that sparked the first living cells.

Deamer's studies have also shown that the kinds of foamy lipids often seen concentrated on lakeshores and seacoasts form sandwiched layers with many

---

*In chemistry, "reducing" refers to substances with high electronegativity that are holding electrons with high chemical potential energy, poised to drive chemical reactions.

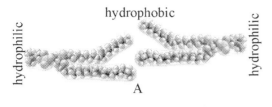

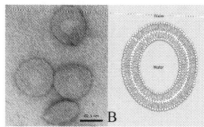

Figure 3.3 Phospholipid molecules as they self-organize in water. *A,* Two paired molecules. The hydrophobic "tails" mingle together, excluding water, as the hydrophilic "heads" are attracted by water on either side. *B,* When many of these molecules coalesce in water or water-based solutions, their polar attraction-repulsion reaction with respect to water induces them to form liposomes—seen in an electron microscope image on the left side, plus a diagram-view on the right. Liposomes are hollow spheres with water inside and out, and are the basic structure of cellular membranes in all living cells. *Credits:* Molecular models from Slideshare.net; Liposome imaging by James Cheetham, Carleton University, CA. Used with permission.

other organic molecules as the foam dries. Then, if rehydrating occurs, the lipids often form complex liposomes, with up to half of the various associated molecules trapped inside, including DNA in some of the experiments (see note 16). Thus the lipid-formed micro-labyrinths that are the membranes surrounding and within living cells likely began as self-coalescing envelopes of hydrophobic lipid molecules, coming together a little like tightly schooling fish. However, if that were the end of it, a phospholipid membrane would be no more conducive to the life of a cell than is the skin of a soap bubble. What matters enormously is this type of membrane's affinity for binding in stable relationships with myriad proteins, another seminal example of self and other generating cooperative emergence. The proteins, crudely at first but with increasing efficiency honed by natural selection, became regulated sites of entry and exit of key material used by and produced by the cells. Eventually, numerous kinds of proteins specialized as intricately shaped pores and channels leading into the cell's interior. Various ions and small molecules such as sugars could then be controlled and guided through their special entry ports, while other substances could be blocked at those sites.

Primitive cells, the first prokaryotes, had arrived; they were composite biochemical replicators with probably crudely regulated metabolism. There was nothing to compete with them, nor, indeed, to stop them except extremely hot or cold surroundings, harsh ultraviolet or stronger (ionizing) radiation, and perhaps extremes on the acid-base scale. From the start these cells may have had the capability, as many bacteria still do, to hunker down in what microbiologists call a cryptobiotic state. Bacteria and several other kinds of organisms do this if environmental conditions become too harsh for ordinary life and survival is threatened. They let go of nearly all their water, but key internal arrays of cooperating protein chains, carbohydrates, and nucleic acids hold together in their vital configurations as the cell shrinks and shrivels into a spore. The external membrane gathers a tough coat of carbohydrate polymer, somewhat like microscopic tree bark, that is incredibly resistant to acids and mechanical forces. Such spores then patiently wait out the environmental crisis until they can again "come alive."

Since the 1980s, experiments in orbit by NASA and European and Russian space agencies have exposed bacterial spores to space for up to several years outside various satellites and the international space station; the spores promptly revived when retrieved and placed in surroundings conducive to their metabolic needs. Most recently, a similar experiment begun in late 2011 took the exposed microbes on a polar orbit that extended well outside the earth's magnetic field that protects against severe radiation. The organisms received radiation doses up to fifteen times the levels experienced in the earlier studies. First results in 2012 indicated that those highly irradiated spores, too, showed no ill effects in terms of revival potential and subsequent metabolism.[18] Laboratory tests confirm the quasi-living entities of bacterial spores readily survive temperatures over two hundred degrees below freezing (of water) and in the hardest vacuum. "Living fossil" spores have been found in deep ocean sediments, having "hibernated" for (it appears, based on how deep in the mud they are buried) millions of years. Bacterial spores are life that is hedging its bets. They exist in a twilight zone between living and nonliving. The simplest form of a bacterial spore may hold similarities to the earliest cell-like bodies to exist on earth.

The Tail That Wagged Itself

Despite its relative simplicity as a life-form, a bacterial cell is still hugely more complex than the putative first self-replicating chemical association in the RNA world. Yet between those two epic markers on life's trajectory, as we have seen in a few among many examples, the cooperative tendencies of biotic polymers, often stimulated by simple properties of their environments, spontaneously shape the stuff of cellular life. Indeed, in chapter 4, we will explore beyond this point and see how whole cells long ago transcended the competitive ethos of self and other to construct myriad progressive biological systems—from advanced cellular life to multicellular organisms to ecosystems, and transformed the planet. The evidence is now overwhelming that cooperation at all levels is in the nature of the universe. There is no scientific reason to doubt that nature's assembly of new, emergent systems with structures of stunning complexity is an unguided, unauthorized evolutionary process—at any level of organization.

It is therefore difficult to understand the pronouncements of a few people, who have been trained as scientists, that some features of cells are irreducibly complex and thus could never have arisen by evolution.[19] Unfortunately there has been a tendency to think of cells and organisms in terms of extreme precision and near-perfection—the Swiss watch model applied to molecular machinery. In the structures and workings of cells—their polymers and organelles, and the ways they behave—we seem to see an unbelievably high level of order. The structures appear to be wrought so precisely and the reactions so fast and crisply tailored between enzyme and substrate. But is it really so? On every level of organization, the concept of life as exhibiting extreme precision—an atomic-clockwork mode of function—appears falsifiable. Life is variable; life is self-organized; it is sloppy and fractally structured in ways that would have greatly disturbed Euclid.

The notion of irreducible complexity in nature is based on faulty reasoning. The whole history of the universe shows us complexity can never be "irreducible." It has always come from some simpler place. Moreover, a living cell seems to work as if cobbled together with odd scraps and parts, and, if creationists were to honestly focus their image of a putative divine designer, they would envision Rube Goldberg or, lately, Wallace and Gromit.

Complexity itself can be studied backward and, less precisely, forward; we can identify the working units of complex systems, often see how they come together, and, in some cases, roughly predict their simplest behaviors or consequences. What quasi-creationists such as biochemist Michael Behe[20] may be thinking of as "irreducible" is emergence. Patterns of complex emergent behavior of systems are unique to particular levels of organization. They are contingent but unpredictable. They are also inevitable, just never in specific terms.

Returning to the proposition of irreducibly complex cellular structures, perhaps it is Behe's type of thinking in the creationist parable of the watchmaker[21] that constitutes the obstacle to understanding complexity whose consequences we see everywhere around us emerging from the bottom up. The parable refers to discovering a watch lying on the ground in some wilderness setting. Clearly any rational mind would infer that it had to have been made by a watchmaker. But for decades it has been clear to molecular biologists that it doesn't take God to fold a protein, and proteins of different kinds spontaneously combine into amazingly complex structures without divine assistance. A philosopher who studies biology carefully might say that neither God nor the devil is going to be found in the structural details of biological systems, but God may be just beginning to evolve in the emergent sequelae.

Those wonderfully intricate but also squishy cellular machines made of proteins, then, process input in the form of typically much simpler chemical materials to produce an output. In the case of a flagellum or muscle fibril, the output is a rotational push or a twitch by a motor protein. This is not surprising, even when it is highly coordinated (although it is unlikely to have started that way). Many proteins change shape naturally; it is simply in their nature, and when various of these molecules form an assemblage that can do something useful for itself there is a tremendous impetus, or intrinsic value, in our universe—that is, there is a selective advantage (in the sense of protecting or expanding the distribution of that assemblage)—to preserve the information needed to replicate the useful function. And once some crudely useful function is preserved, there is the possibility of a push toward greater utility or efficiency through natural selection.

The simplest bacterial flagellum extends outside the cell as a protein filament; it is rotated in a helical pattern, exerting force and moving the cell through water. The filament is powered at its base, where it is attached to a complex of proteins embedded in the membrane of the cell. A so-called rotor complex turns a basal shaft hooked to the filament. This system is powered by a sort of fuel injection, a strong flow of hydrogen ions,* spontaneously derived from water. The ion flow continuously triggers twitching of shape-shifting proteins that induce spin around the circumference of the rotor.[22] Promoting this glorified chemical reaction that mediates mechanical action are nearby ion pumps, channel-shaped proteins mounted in the membrane, that tap the cell's supply of the ubiquitous energy molecule, ATP, to maintain the ion flow.

Proteins controlling ion flows are common in all living cells on Earth. While they and other ultra-miniature organic machines are complex—the rotor assembly of the bacterial flagellum is especially notable—they are all attracted to fit into their finished configurations without external direction. They achieve their complexity spontaneously. The key torque-inducing and basal spinning portions of the bacterial flagellum are also now known to serve an even more vital cellular function (that almost surely arose earlier than flagellar propulsion). Without the attached filament, variations of the shaft and rotor proteins form a sort of chemical mill whose spinning in the membrane energizes the synthesis of ATP itself, the fuel molecule noted earlier, which has been called the universal energy currency of life on Earth.[23]

Behe's pronouncement on the irreducible complexity of the bacterial flagellum seems hard to maintain in light of evidence of highly variable performance by these structures that are made of a variety of structural proteins that react together by changing shape, thus warping, reacting to torque, and spinning the rotor assemblage.[24] As we have noted, proteins are able to bind to each other in more or less precise ways. Thus they often form intricate composite structures: some form enzyme complexes; some bind to membranes; some contract or rotate in response to a nanoscale stimulus; some construct filamentous shapes when they spontaneously join together. Since proteins are commonly irritable—in the sense that they change their shapes slightly in response to light, electricity, ion flow, compression, and contact with chemicals that may be as simple as hydrogen ions—it is not surprising at all that flagella can move. Natural selection has had on the order of three billion years to work on the relative precision in flagellar action that we see today. Even if you argue that the gen-

---

*In some species of bacteria, sodium ion flow sets off the action.

eral mechanism that drives the bacterial flagellum must be conservative—that is, bacteria had achieved reliable locomotion early in their existence—the process of key proteins achieving their appropriate fractal congruence and coordinated reactivity could easily have had many millions of years in the protobiosphere to make connections and fumble their way toward more or less precise mechanical action.

Central to Behe's argument of irreducible complexity is that cell structures can never have had any different function from what we now see. According to this flawed reasoning, these complex organelles must have done what they now do right from the start, and at the same level of efficiency. They could not have arisen by gradual or stepwise assembly proceeding through intermediate structures without discernable function. A major error here comes from not admitting that we very likely fail to recognize possible functions of precursor structures that, in various locations or orientations in cells of the past, would have had uses that were preserved for some time by natural selection. Examples are common in the biology of more familiar organisms: the bones of the mammalian middle ear that transmit the airborne vibrations we hear as sound once formed parts of reptilian jaws. In a different example, highly intricate and highly variable assemblages that are sensitive to electromagnetic energy in animals' eyes trace to simpler complexes of protein and vitamin A that happen to be irritable when photons in the energy range of visible light impact those particular molecular partnerships.

As noted briefly above, the spinning rotor complex and its anchoring proteins of the bacterial flagellum are highly similar to those of ATP synthase, the molecular mill that converts the energy of its spin to the catalytic production of the sparkplug molecule adenosine tri-phosphate (ATP) from its lower-energy ingredients. What earlier, simpler function might have accrued to such a rotating protein system in a formative living membrane? Would it perhaps have entrained a vortex near the membrane, inside or outside, concentrating and channeling small fluid-borne particles—nutrients come to mind—in a microscopic whirlpool that would draw such particles close to the cell for absorption? So far we do not know. However, that's a far cry from asserting there would have been no function possible, not ever. Most scientists, ever hopeful of discovery and new insight, would not draw conclusions from such anti-speculation, which, to a theologian, might resemble a sin of despair.

Michael Behe and the quasi-creationists seek reasons to believe in a divine designer, the author(itarian) of the just-so universe. It is true that we have not yet discovered a function or functions for the earliest clunky spinning proteins in an ancient cell's membrane, but it is more positive scientifically to speculate that logical putative functions may turn up and even be testable. The fine details of ATP synthase and flagellar motors are increasingly coming into focus. That the rotor assembly could have easily hooked up to a thin molecular filament is a no-brainer. There's all the logic in biochemistry in that step, given the innate tendency of proteins to associate.

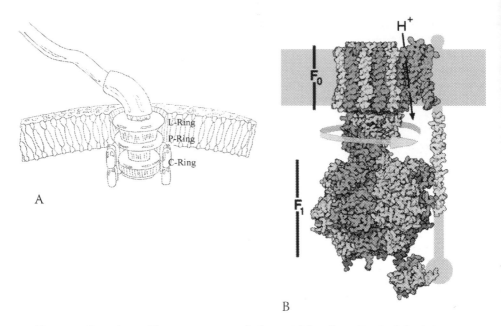

A

B

Figure 3.4 Two views of key components of a bacterial flagellum: On the left, *A*, is a mechanical model, popular in publications of "creation science," showing a bacterial flagellum with its motor assembly embedded in the cell's plasma (boundary) membrane; on the right, *B*, is a molecular model of cooperating, self-organizing, shape-shifting proteins in the embedded base of the flagellum that imparts motion to the flagellar filament (not shown). The cell's membrane in the right-hand image is indicated by the gray zone. The labels "F_0" and "F_1" indicate key energizing and structural protein complexes. "H^+" indicates hydrogen ion flow. The mechanical rendition (left), redrawn from a popular online model, incorrectly depicts the membrane's phospholipid "tails" as continuous across the membrane. *Source:* Flagellar motor protein assembly (right) from Wikipedia, Wikimedia images, public domain. *Credit:* Drawing (left) by Susan Culliney.

The history of progressive emergence in our universe depends on self-organizing cooperation that breeds far-reaching shifts in the course of evolution, and if one were searching for divine attributes that reflect nature, cooperation would probably top the list. It is a cardinal principle at every level of chemistry and biology, and as we investigate the more complex landscapes of those scientific ways of knowing about our surroundings and ourselves, certain emergent manifestations stand out above the rest. They lead the way toward new peaks of diversity in nature or to bursts of evolutionary change; such is the extensive influence, or weighted consequences, accruing to systems in which they dwell. Metaphorically such entities appear ordaining; they often act like tricksters in shaking up their systems. In their subsequent influence in the world we might call them keystones, a term applied in ecology (as discussed in the next chapter).

In biochemistry, keystone emergents include ribozymes and enzyme proteins as exemplified above. The chlorophyll molecule is another profound example that proliferated in emergent blue-green bacteria across the world's seas over three billion years ago, eventually triggering a revolutionary change in the planet's atmospheric and oceanic chemistry. Without that change, we and nearly all the rest of the visible biosphere would not be here.

Ecology Emergent

> I had lifted up a fistful of that ground. I held it while that wild flight of
> south-bound warblers hurtled over me into the oncoming dark. There
> went phosphorus, there went iron, there went carbon, there beat the
> calcium in those hurrying wings. . . . I watched that incredible miracle
> speeding past. It ran by some true compass. . . . It cried its individual
> ecstasies into the air. . . . It swerved like a single body, it knew itself and,
> lonely, it bunched close in the racing darkness.
>
> —Loren Eiseley, *The Immense Journey*

The emergence of cellular life from the world of complex carbon-based chemistry appears to have happened only once in the primordial dreamtime of planet Earth. Scientists base this conjecture on a number of virtually universal distributions of chemical structures and processes across the spectrum of living organisms. Despite their perhaps tenuous hold on life, the first cells possessed the keys to the opening of new potential for matter and energy—the capabilities of self-replication, controlled energy transduction, directed locomotion, and the regulation of an internal environment. Out of this cellular Big Bang arose a totally new force field on planet Earth, superimposed over the physical, chemical, and geological but with tendrils interacting with all of those realms. It was the beginning of the biosphere. Life pervaded and began to transform the lithosphere, hydrosphere, and atmosphere.

Ever since life's debut on the Earth, biotic evolution has been a near-balancing act. On virtually every level, competition and cooperation, integrity and intimacy, shifting endlessly between foreground and background, have tugged and teased evolving systems as they have wobbled through time along the edge of chaos.

Emerging out of cellular forms that we recognize as the most basic templates of life are composite organisms, made of cooperating cells that achieve progressively greater sophistication in a continuum that leads from initial mergers of diverse prokaryotes in endosymbiosis (see next section) through cellular colonies, small and large, to the most highly integrated bodies of animals. There appears to be no way to discover or capture the entire kaleidoscopic succession.

Most of the evolutionary drama played in long-ago ecological theaters.* The fossil record is suggestive but can never be completely revealing. In many cases, traces of the actual links, as real organisms, have vanished, although more information may yet be glimpsed in the membrane captures of endosymbiotic events and in genes from the likes of contemporary sponges and cnidarians, echinoderms, and chordates that will enable further reconstruction of some of our ghostly common ancestors. But through all the long burgeoning of life on Earth, the building of structure in the biosphere—leading to astounding paragons of emergence beyond the simple primordial soup of cells in the sea—has strongly depended on the tendency of life to form cooperative associations. This process has resulted in progressive complexity, as life mutated and differentiated into its myriad manifestations. And again the cooperative impetus, the tendency to attract the other and develop mutually supportive functionality, has been confirmed by natural selection throughout biospheric time.

However, we do know already that the road through the rising terrain of biological complexity, including symbioses of cells and organisms, has taken a very winding and devious route. This understanding is strictly modern, in contrast with prevailing evolutionary views through much of the twentieth century, during which the reigning paradigms represented neat, steady progress toward individual, complex, integrated life-forms: the so-called higher animals and higher plants. This view is now supplanted by a much clearer understanding of a general adaptive evolutionary process that appears nonlinear and haltingly progressive, in which complexity is not straightforward, though expected.

Evolution of the Earth's overall biosphere has proceeded in two main directions, one dependent on and following the other, and then interweaving with prodigies of emergence. The first to get started was a tremendously radiating spectrum of microbial life. And then, after a longish interval, a series of seminal mergers of very different types of these cells produced a new energetic wave of diverse living forms.[1] This became a hugely emergent phenomenon that largely focused in a deceptively narrow ascending trajectory of multicellular organic capability to amplify *structural* diversity of the biosphere, building ecosystems far beyond microbial levels. And at every juncture, the evidence of cooperation as the source of progressive complexity in the biosphere is compelling.

Life after the Beginning

Thus, life first burgeoned into a vast underworld of tiny, primitive-appearing cells, having an apparent simplicity of structure but with immensely diverse biochemistry. This is the prokaryotic, or microbial, biosphere, a galaxy of living systems that has hosted the most prolific adaptive radiation of proteins, lipids, and their hybrids. Until very recently, the degree of variability of this world has been all but invisible to us. Now, the latest methods of molecular biology—the

---

*Homage to G. Evelyn Hutchinson for his insights into the evolutionary drama.

ability to rapidly decode sequences of DNA—have begun to reveal that our planet is still profoundly dominated by millions of microbial types. Hundreds of kinds have already shown themselves to be at least as different from each other in their cellular chemistry as plants are from animals. In these microorganisms have evolved processes feeding their metabolisms that are able to alter the great

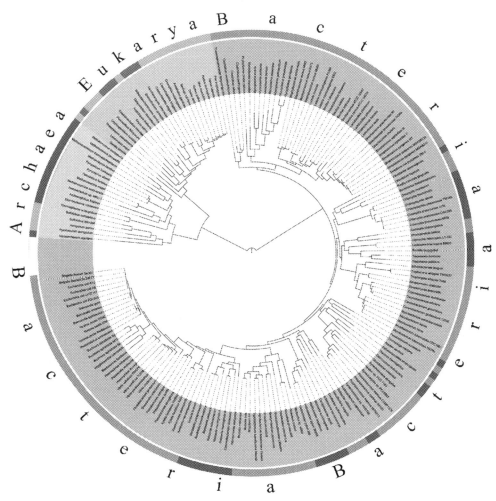

Figure 4.1 The biological multiverse. Diversification of life in radiating patterns from its cellular inception after more than 3.7 billion years. Three large groups of organisms known as domains of life, the Bacteria, Archaea, and Eukarya, emerged as the earliest surviving descendants from a common ancestor. The first two are considered to be simpler life forms at the cellular level, and both are referred to as microbes. Recent discoveries of new microbial variants in the sea, especially bacteria, will hugely expand the catalogue of biodiversity. *Source:* Public domain image from Interactive Tree of Life (ITOL), http://bioinformatics.ca/links_directory/tool/12182/itol, via Wikipedia: https://en.wikipedia.org/wiki/Phylogenetic_tree#/media/File:Tree_of_life_SVG.svg.

geochemical cycles of the oceans—and by extension, the atmosphere—at times triggering global shifts of carbon, oxygen, and nitrogen. The almost unbelievable planetary abundance of the microbes is revealed by recent widespread surveys estimating that, on the average, over 150,000 of their cells, distributed among at least a thousand species, coexist in every cubic centimeter of sea water.[2] There are hints that some of their genes may trace back to early ages of the biosphere. And just as fast as they are being discovered, these organisms are being found to harbor genetic codes for biochemicals that are entirely new to science and whose functions as enzymes, industrially useful polymers, new pharmaceuticals, and the like may prove enormously valuable. The exploration of this hitherto hidden world has just begun.

Between 2003 and 2008, one of the pioneering explorers of the microbial biospheric frontier, J. Craig Venter,* led a globe-circling oceanic expedition that filtered bacteria and archaea from hundreds of widespread seawater samples. Genes of the microbes were sequenced using a rapid method to compare numerous comparable strands of DNA—revealing overall genomic diversity and, in turn, indicating protein products of many of the genes. The first leg of Venter's expedition aboard a specially outfitted sailing vessel, *Sorcerer II*, between Halifax, Nova Scotia, and the Panama Canal, produced an apparent doubling of the known major protein varieties, accounting for up to several million new enzymes.[3]

Evidence of gene-trading among diverse marine microbes was conspicuous in the samples. The traders are often viruses that pick up variable sections of bacterial DNA from one cell and carry it into another during reinfection. This DNA dispersal commonly crosses wide taxonomic boundaries; when viruses carry nucleic acid of a particular host and insert it into the genome of a new, different organism, they blur the identities of self and other. In the sea, viruses having microbial hosts vary enormously and occur everywhere, and they actively jump far wider taxonomic gaps than does, say, a bird flu virus that can infect a mammal.

The overall genetic sequences in the DNA sampled from the top one thousand meters of seawater along *Sorcerer*'s Atlantic cruise track changed by about 80 percent every two hundred miles, and that rising curve of new microbial types showed no signs of leveling off as the search moved into the Pacific. In Venter's words, "We have not understood much about our own planet and our own environment. We've been missing as much as 99 percent of the life forms and biology out there."[4]

Beyond microbes, the second striking pattern (amid the heretofore invisible prokaryotic dominance) has been the radiation of the eukaryotes as the most conspicuous domain of life. This second pattern ultimately emerged as a soaring pinnacle of structural variability and wondrously complex life-forms—ferns and trees, corals and squids, dinosaurs and whales—but with biochemical

---

*Venter is best known for leading the private research group that in 2001 published the first fully sequenced human genome. See J. C. Venter et al., "The Sequence of the Human Genome," *Science* 291, no. 5507 (2001): 1304–1351.

diversity far less extravagant than in bacteria and archaea. The evolutionary lineage that would spawn such palpable splendor in the biosphere initially depended on prokaryotic unions (providing compelling evidence of the cooperative constant) via the phenomenon now known as endosymbiosis. New composite cells, stem ancestors of the Eukarya, then began to explore their way into the dazzling macro- and mega-scale morphological, physiological, and behavioral achievements of life within the waters, over the landmasses, and across the skies of the earth.

Hybrid Vigor

One of the most significant realizations of twentieth-century biology, right up there next to Watson and Crick's monumental discovery, was the theory of endosymbiosis: the origin of the cells of complex organisms, the Eukarya. This theory sprang largely from the mind of Lynn Margulis,[5] although earlier observations on the idea by German, Russian, and American scientists, notably Andreas Schimper, Konstantin Merezhkovsky, and Ivan Wallin, had been around since the late nineteenth century. Building on numerous, increasingly sophisticated cytological and biochemical observations that had accumulated without focus, and with her own discoveries and insightful interpretations, Margulis proceeded to clinch the compelling case for the origin of eukaryotic cells from mergers of different early kinds of bacteria. The most prominent examples were those that, after becoming enclosed within a perhaps sluggish host microbe, turned into the vital eukaryotic cell organelles known as mitochondria and chloroplasts. Today no known prokaryotes have these sorts of organelles. That mitochondria, the strange but ubiquitous ovoid organic energizers (chloroplasts are the solar-charged version) once were free bacteria themselves is now clear from several lines of evidence.[6] Those findings include the fact that both of these intracellular bodies possess their own genes, in bacterial arrays on single, small, looped chromosomes of the type possessed by all bacteria.

In addition to acquiring their symbiotic energizers, nascent eukaryotic cells reorganized their genetic material into individual chromosomal strips and enclosed them within a doubled lipid-bilayer membrane (see also chapters 3 and 5 regarding the significance of those cellular features). The arrival of the eukaryotes became the most far-reaching milestone of biotic evolution stemming from the cooperative principle at the cellular level, an intimacy of self and other with extraordinary potential. No Darwinian gradualism here: in a few profoundly punctuating moments in ancient seas, individual bacteria-like cells of radically different types melded into new emergent life that would transform the biosphere.

Exactly what triggered the events that created the ultimate ancestors of plants and animals, among numerous other "advanced" life-forms, is unknown. Simple suggestions include unconsummated predation or parasitism. What is increasingly known is that endosymbiosis was not a one-time, or two-time, wonder. It has happened many times, and to many various bacteria, and has even attracted later eukaryotic cellular mergers of distinctly different types. The known telltale traces involve multiple layers of concentric internal cell membranes in descendant species today. Myriad early cells very likely gravitated toward partnerships with many benign outcomes, but as expected from the power curve of emergent events, only rarely did such a tiny organic ripple amplify to become a swell of great influence.[7]

In keeping with the fractal principle of self-similarity across scale, it might be fruitful to search for parallels between self-organizing cellular structure and function and larger-scale patterns that can be understood as self-organizing between and among whole multicellular organisms and in ecosystems. Basic patterns governing the evolution of those emergent systems seem highly likely to resemble those leading to spontaneous molecular cooperation in any type of cell.

Now, many researchers, taking cues and clues from Lynn Margulis, have uncovered a vast province of *extracellular* bacterial mutualisms—with prokaryotic partners and with animals and plants. Thus, within a particular "habitat"— whether in a cell or in a complex, multicellular organism or in a forest or on the bottom of the sea—intricate structure and function appear as interacting dissimilar partners adapt to the local physical-chemical milieu and to each other. This happens in the absence of any imposed design from outside the system; indeed, the system itself changes, becoming more powerful in the sense of triggering changes, from local to global, and regulating what happens within and around the system—storing and organizing greater amounts of matter, energy, and information, and functioning more smoothly as time goes by. Living systems introduce new patterns of order in nature; they become *self-invested* with the principle of creation itself.

The Life Gregarious

A marine microbiologist dips a sterile glass slide into the sea and within a few seconds a startling self-organizing process begins—the formation of a microbial community called a biofilm. The first species to attach to any freshly exposed surface in this situation are several kinds of bacteria distributed throughout the world ocean. Left undisturbed, within minutes they begin to divide and form colonies, spreading out on the glass. Soon other life-forms arrive: new bacteria, archaea, and eukaryotic cells such as diatoms. They arrange themselves in characteristic patterns over the surface; the organizing principles oscillate between competition and cooperation; however, complexity inevitably gains ground as cooperation rises and competition devolves to tolerance and truce. Emergence happens; structure in a physical sense depends on specific membership in the community. Even in the simplest of biofilms, experiments involving two cooperating bacterial species have revealed the emergence of a particular spatial orientation—one self facilitating the growth pattern of the other in intimate choreography on the surface where the cells settle and proliferate.[8] With numerous species that eventually participate in a biofilm, a kind of layered micro-ecosystem develops: a microbial canopy with subsidiary tiers, on down to the basal colonies. Gradients of sunlight and oxygen may be present. On surfaces considerably larger than a microscope slide, the initial biofilm that may be highly complex in its microbial composition ultimately attracts bigger crea-tures, spores of seaweeds and larvae of marine invertebrate animals, and across much of the Earth, the biodiversity of the sea floor rises toward levels found in tropical rainforests.

Such cooperative interaction in nature progressively raises the complexity of the world. Intimacy of self and other then is tested by natural selection for its adaptive value. And potentially a brand new system appears with emergent power and utility, sometimes even breaking through to generate a new prospect for life, a new ecological niche that was never there before.

Examples such as biofilms indicate that life is primed for cooperation. Of course, as we have noted, the edges of this great basin of attraction, the bio-sphere, are always turbulent, full of surprises. Sometimes mutualism on one level comes at the expense of another—as in the case of well-known biofilms in the form of plaque deposits in the human mouth. In such cases, a living host—in this case, the human body—becomes an ecosystem with cooperation on one level and exploitative interactions on another. But in numerous examples, re-searchers have begun to discover evidence that pathogens commonly have the tendency to evolve toward mutually beneficial relationships with their hosts.

It is easy to understand how disease-causing organisms—from viruses to animal parasites—end up being selected in Darwinian fashion for decreasing virulence over time. Any pathogen is a parasite, and for such an organism to be successful in the context of natural selection is to maximize its reproduction. A very high degree of virulence then is usually self-defeating for the pathogen. Suc-cess often means evolving toward a less-than-lethal effect on a host. However, until Margulis pointed to the profound cooperative nature of the biosphere, the focus of biologists mainly emphasized the antagonistic nature of bacteria–host relationships. The workings of animals' immune systems, which so strongly de-fine self and other in biochemical terms, and the study of diseases of agricul-tural crop plants thoroughly dominated the field of thought.

When animals and plants arose, those eukaryotic forms became new habi-tats for microbes, and so organisms routinely became ecosystems. Myriad mi-crobes colonized multicellular creatures and became permanent residents. In animals the great microbial attractor was the gut, where the symbionts estab-lished their filmic forests and prairies in cooperation with their particular host. The microbes benefited from the particular diet of the host, which concentrated a range of food material that varied with the host species, thereby attracting different suites of microbes. The host made use of vital products of the symbi-onts' metabolism: for example, in animals as different as mammals and mol-lusks, using enzymes to break down tough-to-digest material such as cellulose into its component sugar molecules. Other symbiotic bacteria contribute various vitamins that are essential to animals' health. Some specialized bacteria in these strong mutualistic symbioses also produce essential amino acids for animals and others generate nitrogen fertilizer for plants—and the list of microbial ser-vices goes on.[9] The vitality of animal-microbial ecosystems, exemplified in the gut-as-habitat in omnivorous mammals, is revealed by bacterial cells outnumbering those of the host at least tenfold. Such internal cooperative assemblages are now well known in vertebrates, insects, mollusks, and annelids, but to echo J. Craig Venter, the exploration of the diversity of these inner microbial worlds of eukaryotes, like the oceanic microbial ecosystems, is in its infancy.

As noted above, our understanding of the universe of living partnerships between microbes and animal hosts was impeded by the long-term focus in biology on the integrity-preserving immune systems of complex eukaryotes as the rigid enforcers of distinctions between self and other. A new, more nuanced paradigm has emerged that acknowledges ways in which animal cells and tissues respond to contact with alien microorganisms in complicated chemical communications that indicate compatibility versus threat. And it is ever more obvious that both animals and plants respond to signals of compatibility, welcoming myriad creatures into their bodies where mutual transformations then may proceed, progressively blurring the identities of self and other.

Light in Dark Waters

Sometimes microorganisms form one-on-one partnerships with eukaryotes, and emergent propensities of those symbioses may release truly global manifestations. Such is the case with bacterial bioluminescence in the sea. A large number of organisms, especially certain marine bacteria, produce biological light, which is a highly specialized biochemical property within cells. The most common light-generating process is a reaction of a molecule called luciferin. This reaction is triggered by an enzyme known as luciferase. The luciferase gene and the biochemical pathway for manufacturing luciferin are widespread but probably bacterial in origin. The luciferin-luciferase reaction is unlikely to have evolved independently on multiple occasions in very different creatures. If that is the case, then the sharing of the genes over eons by luminescent bacteria with many other forms of life—via gene swapping mechanisms such as viral infection—has released an emergent transduction of energy to enhance the complexity of the biosphere. The patterns of behavior arising from this phenomenon represent a kind of informational surge—in terms of species interactions and communications—using cold light to attract prey or a mate, deceive a predator, or find prey in the deep sea. Gradually, through deep time, dissimilar creatures have created ghostly glows and bright semaphore flashes, lines of tiny shining portholes and baroque patterns of scintillations weaving through the primordial darkness of the world.[10]

Among the more startling examples of bioluminescence are actual operating partnerships between bacteria, such as certain *Vibrio* species* and *Photobacterium*, with animals, including various squids and fishes. One of the best-studied

---

*To many biomedical scientists, veterinary pathologists, and the like, *Vibrio* species and related bacteria are better known for causing serious disease in animals, including human cholera. *Vibrio fischeri* and others involved in bioluminescent partnerships do not appear pathogenic to their normal hosts.

Figure 4.2 The squid *Euprymna scolopes*, after hatching from its egg capsule, begins a symbiotic partnership with the light-emitting bacterium *Vibrio fischeri*. Each partner has affected the evolution of the other to their mutual benefit. *Credit:* Image provided by Dr. Margaret McFall-Ngai, University of Hawai'i. Used with permission.

of these illuminating symbioses is between the Hawaiian bobtail squid (scientifically, *Euprymna scolopes*) and the bacterium *Vibrio fischeri*. The squid is a small species that spends the day hiding in sand close to shore around the islands; the bacterium occurs free in seawater, as well as in extreme concentrations in small twin-lobed pouches, suspended low in the hollow space, called the mantle cavity, that houses the squid's major mass of vital organs.

Picture the structure of the light organ in this squid as an array of organic components crudely analogous to those of the front end of a flashlight that is adjusted to shine with a wide beam. The light-emitting vibrio grow densely amid clusters of projecting filaments of squid tissue. These filaments are centered in a cuplike reflector, lined with an iridescent organic film that reflects the glow from the closely packed bacteria through a diffusing lens with an iris-like fringe that modulates brightness. The whole unit has also been likened to an eye—but one from which light shines outward rather than being taken in.[11] The wide cone of light emerging from the lens causes the translucent mantle tissue of the squid's belly to glow softly like a Chinese lantern.

Inoculation of the light organ with the light-emitting microbes occurs shortly after the baby squids hatch out of their protective embryonic capsules (eggshells) on the reef. Hatching occurs always at night, and it is possible that the tiny animals (about one-quarter inch long) with their large eyes can detect, at near-microscopic range, the bacterial sparks bathing them. However, those vibrio the little squids seek probably always constitute less than 0.1 percent of the open-water marine bacteria. Nevertheless, a reflex in the squid rapidly secretes a mucus attractive to the bacteria. The secretions occur on the host's skin at the sites where recruitment of the vibrio will enable their absorption. Once inside the ducts leading to the tissues of the embryonic light organ, it appears that bacteria are further selected by releases of chemicals that eliminate bacteria other than *V. fischeri.*

Researchers believe that the inoculation is accomplished during minutes to hours on the first night that the new generation of squids swims free in the near-shore waters. In the mantle (respiratory) chamber of the squid, concentrating vibrio into the light organ appears to be an active process by the animal. Tiny cellular brooms (clusters of cilia) sweep the microbes into small pores leading to the incubation surfaces of the eyelike cup, where the bacteria are attracted to settle, multiply by millions, and switch on living lanterns.

Unlike many bioluminescent animals, the bobtail squid possesses only a single light organ, which takes its initial shape during embryonic development. Some of the most intriguing studies of this symbiosis have revealed the intimacy of the association in terms of each partner inducing anatomical changes in the other. On the part of the squid, the shaping of the light organ from its embryonic form to the complete functional structure takes about three weeks and will not occur in the absence of the vibrio partner. The reflector and lens, for example, develop from a rudimentary state and seem to depend on chemical communication by the initial colonizing bioluminescent microbes setting off a chain of events in the squid.[12]

In the case of the bacteria, within hours after implanting in the embryonic light organ, they lose the close cluster of multiple flagella *Vibrio* species use for swimming in their free, wild state in the sea. The bacterial cells also become much smaller as their populations proliferate like banks of microscopic Christmas-tree lights amid the dense shrubbery of compatible tissue provided by the host. One further surprise in this symbiosis is that the squid, each morning around dawn, eliminates 90 to 95 percent of its vibrio symbionts back into the ocean. Then, through the day, those that are left inside divide rapidly and repopulate the light organ to shine at full strength when the squid again patrols the shallows by night.

Eons ago, the advent of the first bacteria to produce biological light changed forever the nature of the nighttime sea, and especially the abyssal ocean where sunlight never shines through a vast swath of Earth's biosphere. Partnerships between luminous bacteria and animals brightened the depths further. Finally, genetic acquisitions from the bacteria conferred direct (intrinsic) bioluminescence in a variety of marine animals. Those achieving self-generated illumination are usually thought to be more highly evolved than those embodying functioning partnerships, such as between *V. fischeri* and the bobtail squid, in which light-manipulating mechanisms sometimes evoke Rube Goldberg–style contingencies. Reaching intrinsic status seems to indicate a progression through intimacy toward integrity. However, it seems likely that across the bioluminescent spectrum, partnerships initially switched on the lights of various types of marine animals. Hints of the progression are indicated in that luciferase genes—and others that regulate luciferase expression in eukaryotes such as the protistan dinoflagellates—most often lack segments, called introns,[13] that distinguish eukaryotic genes from bacterial genes (which generally lack the segments). It is also possible that some hosts absorbed their light-emitting prokaryotic partners into endosymbiotic associations and ultimately maintained only rudiments containing the basic system of illumination. Such cases might still be detectable using electron microscopy. And in any case, the acquisition of intrinsic eukaryotic bioluminescent capability would still reflect intimacy on the level of bacterial genes transferred to the host, a legacy of prior cooperation shining across emergent levels of organization. New luminous beings—protists, fishes, shrimps, squids—progressively began lighting their ways in the world.

Constructions Visible from the Moon

The ocean was a cradle of symbiotic associations, from ancient days, of polymers in the surface microlayer. Long after that early experimentation involving insensate attractions among large organic molecules, microbes, followed by ever-more-complex creatures, began to test mutual compatibilities. Among the most spectacularly emergent at the organismal level was an entirely eukaryotic paragon of partnership—the mingling of many species of coral animals* with a type of algae, a photosynthesizing organism called a zooxanthella. These single-celled algae are taken in by the corals or, alternatively, are said to infect their coral hosts (the active roles of the partners as they form this association are not quite clear). Perhaps, as in the bobtail squid's relationship with its bioluminescent microbes, both partners actively engage in setting up housekeeping together.

The algal cells resemble tiny golden balls about eight-thousandths of a millimeter in diameter. They are packed with chloroplasts and, basking within the translucent tissues of corals in the sunlit shallow waters of the reef, they produce sugar, the organic product of photosynthesis. Zooxanthellae make so much sugar—beyond their own needs for energy—that much of it leaks out to their host. And the corals that house zooxanthellae gain a second big advantage—one that boosts the growth of their stony skeleton. In performing photosynthesis, all plantlike organisms, including algae, absorb carbon dioxide, which is the carbon source for their synthesis of sugar. Remove CO_2 from water, and you reduce its acidity. In reef-forming corals, this is a substantial effect. The crystals of aragonite (a kind of limestone) that are organized by coral tissues to form the underlying skeleton are greatly affected by slight changes of acidity. Limestone dissolves in weakly acidic solutions. By scrubbing the CO_2 and thus the carbonic acid, zooxanthellae strongly promote precipitation of limestone over dissolution,

---

*The animals discussed here are specifically hermatypic scleractinian anthozoans, known also as reef-building stony corals.

and those types of corals with zooxanthellae symbionts, called hermatypic corals, are the principal reef-builders, owing to the high rate of their skeletal limestone production.[14]

The coral-algal symbiosis of the hermatypic association is exceptionally tight. The intimacy is so strong that some biologists consider the partnership as an integrated organism in its own right. Yin to yang, intimacy evolves to integrity, as realized in the partnership. Neither partner is able to thrive independently for more than a few days to weeks. Environmental disturbance, especially slight warming of seawater (often no more than two or three degrees Celsius) above the normal summer maximum temperature for a given species of coral, causes the coral to expel its zooxanthellae, a reaction known as coral bleaching. Excessive turbidity from silt in the water also may induce bleaching. Instead of the typically healthy beige to golden-brown color exhibited by most hermatypic species—their color in a thriving state most commonly reflects the algal hue—the bleached coral goes dead white. Without zooxanthellae, the lack of pigment in the limestone skeleton is thus revealed under the ultrathin, translucent animal tissue. Coral bleaching is usually fatal. Only occasionally, with the reestablishment of cooler or clearer water, does the coral acquire new zooxanthellae and a new lease on life. Bleaching episodes often kill corals across many square miles of reef, and this threat to the most diverse ecosystem in the ocean is increasing with global warming.

Hermatypic corals are the keystone species (also known as foundation species) that have led to the emergence of one of earth's richest communities and ecosystems.* The rapid growth of stony skeleton of these corals, facilitated by the zooxanthellae, produces the so-called framework of the reef: limestone slabs and domes, plates and candelabra forms, all in a seemingly chaotic but biologically intricate array across the shallow tropical seafloor near an island or continental shore. In and around this self-organizing chockablock rockery, many other kinds of organisms find shelter, food, and opportunities to reproduce. Over time many of these creatures, from cyanobacteria (blue-green algae) and various seaweeds to foraminiferans, clams and snails, urchins and sponges, and many others also contribute to the physical construction of the reef. Their contribution comes in the form of sediment—sand and silt produced from erosion of their skeletons—and the sediment, itself mostly particles of limestone, often adds up to at least as much mass as the large chunks of coral framework. It drifts like snow into the hollows, crevices, and rugosities of the seabed. It compacts as it deepens and eventually sets like cement to lithify the whole bottom

---

*A biological community can be defined as a group of organisms of different kinds that live in a particular habitat and interact in important ways: as competitors, cooperators, photosynthetic producers, prey and predators, and so on. An ecosystem consists of a particular type of community, such as a forest, grassland, tundra, or coral reef, with all of its biological interactions, plus the physical, chemical, geological, meteorological, or aqueous relationships of the environment with the biota.

deposit into a rough limestone conglomerate a few feet below the surface of the living reef.[15]

Some coral reefs whose surfaces we snorkel over today are ancient. When sea level slowly rises or the sea floor subsides, reefs grow upward. In some places they become hundreds to thousands of feet thick. Always, the entire reef mimics a typical coral colony in that, overwhelmingly, living matter is concentrated on the reef's surface, but the whole stony mass, in depth, preserves its skeletal history through long spans of geological time. Alongside these massive-yet-intricate, communally grown earthworks, human constructions seem evanescent. And even those deeply buried former coral seascapes may come to light in a new age: if they are stranded on future continental margins by tectonic upheaval into high mountains, then erosion gradually uncovers their contours once more, with stories of lost horizons to be read by paleontologists and Zen practitioners.

Ecopoesis: Burgeoning on Earth

From biofilms to coral reefs to forests, grasslands, and arctic tundra to the vast oceanic pelagic and the deep sea floor—throughout the planetary biosphere—life self-assembles into natural communities. Each community develops a particular ecology-at-large, governed by an interplay of physical and biological forces. This tends to determine the member species, together with their distribution and abundance, and inevitably a biotic community transforms what was there before in the physical space it comes to occupy. One primordial example we like involves an initially sterile geothermal spring with a range of chemical variability: sulfur emissions, carbon dioxide, various dissolved salts, and so on that are simply outputs of the usual raw hydrolithic processes of our planet.

Over time, affinitive microbes of various kinds find such a place via atmospheric spore dispersal and begin to colonize it; what will happen? Of course they will multiply, at rates that may be different among the species represented. Different species will seek and grow on or in different physical substrates of the spring—exposed rock surfaces, cracks in the bottom away from light, or in mineral mud, for starters. Some may prefer the shallow rim of the pool; others, the deeper zone. Some may prove inimical to others, competing for surface area in which to grow, or for prevailing concentrations of nutrients. All these species will spontaneously generate increasingly complicated patterns of spatial distribution. Cooperation will also happen, sometimes as cooperation that depends on others' competition. There may well develop gradients in population density of various species along correlating gradients in temperature, nutrient concentrations, or light intensity. With time for evolution via mutation, genetic recombination, natural selection, genetic drift, and symbiosis, ever-more-intricate patterns and interactions will begin to appear. Like the face of the Cheshire cat, the structure of a particular microbial ecosystem, an assemblage of interacting species with the propensity to maintain and diversify the assemblage itself, will materialize out of what had been a lifeless pool of water and minerals. Indeed, the very physics and chemistry of the original space—the hot spring—will change, owing to the complexity induced by the suite of organisms interacting with many features of the physical habitat as well as with each other.

The ecology of a hot spring is a very simple example, just a small step beyond a biofilm. In the biological history of our planet, the greater the complexity and diversity of the life-forms, the greater the departure of the resulting ecosystem they concoct from the physical-chemical nature of any kind of raw habitat, but an observer with the simplest scientific tools, such as a microscope, can verify that it is clearly all self-organizing. It takes no guiding hand or divine designer to place the organisms together, just so, in soils, or lakes, or forest canopies.

Ecologists have found that the trajectory of a community through time commonly unfolds in a developmental pattern that—appearing to follow principles of deterministic chaos—gradually builds complexity through a process called *ecological succession*.[16] For any given community, succession is classically considered to pass through a number of stages called seral stages, or seres, and the theoretical endpoint is a final stage of sustained equilibrium and maximal complexity called the climax, in which a roughly definitive membership of species persists in a specific geographic setting: say, a forest in Ontario, or a coral reef fringing part of Maui. However, ecological researchers have realized for some time that seral stages are illusory in a progressive but turbulent fluid process, and a climax is rarely if ever approached.

What happens is that species replace each other in "assortative" waves, as first identified by Edward O. Wilson,[17] spreading and interacting in more or less predictable patterns as a community develops on a landscape or seascape. In its primary form, succession begins with a so-called pioneer assemblage, made up of the first interacting denizens in raw geophysical habitat. The pioneering species typically form a consistent group of colonizers of their particular habitat category and planetary locale; eventually they give way to others, as time passes, and largely different species compositions prevail in the community's later seral development. All the while the community burgeons in diversity, net productivity, and other measures of ecological complexity.

Simply by following built-in natural rules and properties of their kind, the species that constitute biotic communities collectively sort themselves out; over time and space in their habitat they find each other, and, very like the components of a flagellum or a living cell itself, they construct a fantastically baroque, living, evolving mechanism that progressively transforms part of the world. In every case, the emergence of order from chaos seems improbable, but we now realize that on every level it is inevitable, with the most intricate "squishy" living mechanism—an ecosystem—taking spontaneous shape and exhibiting amazing properties of sustainability, energy metabolism, and sophisticated storage and processing of information. And this gives us cause for hope. Even the Earth as a whole, in James Lovelock's Gaia hypothesis,[18] has been construed by some to be such an evolved mechanism, (albeit a very sloppy one), with the web of life that has woven a vibrant fabric within its waters and across its wrinkled lands, integrating vast and progressive change in its physical-chemical identity as a planet via increasing intimacy with its biota.

In the next chapter, we will examine the cooperation of self and other in one of the greatest biological mysteries: the evolution of sex. As one of the most powerful attractors in the living world, sex reduces the integrity of participants in favor of intimacy but paradoxically (and immensely) enhances the potential for emergence in the biosphere. To a Daoist biologist, sex as participation in determining traits of future generations represents one of the keystones in the way of the living world, providing a bridge to emergence in the biosphere. Buddhists, too, see the state of interbeing in the manifestations of sex as the most natural pattern in the world. However, biologists, caught up in reductionist models of understanding nature, have had greater difficulties trying to understand the origins of sex and what has driven its evolution, nearly universally, in complex organisms.

Intimate Ark
Sex and Emergence

> We are composites of many different legacies, put together from
> leftovers in an evolutionary process that has been going on for billions
> of years. Even the endorphins that made my labor pains tolerable came
> from molecules still shared with earthworms.
>
> —Sarah Blaffer Hrdy, *Mother Nature*

Among the greatest cooperative milestones of biotic evolution that released a virtually unbounded world of complexity, particularly conspicuous among eukaryotic organisms, was the evolution of sex. In sex, each individual of a mating pair contributes part of its genetic makeup (genome) to offspring—always cells are the seminal agents of the genetic contribution from a self—that participate in an emergent new generation. Thus each self, upon engaging in sex, abandons a substantial portion of its integrity and weaves together a fractal interface with a partner. Throughout the sexual world, *self* seeks a profound intimacy with *nonself*.

However, unlike the merging of self and other in the striking cooperative endosymbiotic adaptations that brought complex (eukaryotic) cells into the world (see chapter 4), the earliest unions of cells that led to sex were not an obvious step toward complexity. The first organisms groping toward sexual expression were almost certainly prokaryotes, but their participation in genetic intercourse has remained incomplete and relatively imprecise (see below). The eukaryotes, however—from tiny protistans and filamentous fungi to the many various forms of complex animals and plants—evolved a process of amazing molecular precision and generated hugely diverse manifestations of emergent anatomy, physiology, and behavior in the service of sexual reproduction.

Bacterial Sex: A Simple Affair

In the roiling waters that formed the cradle of primordial life, sexuality took time to ripen into a state of maturity and richness. Initially, an ebb and flow between expressions of integrity and intimacy played out at molecular and

cellular levels of organization. Even in the RNA world, occasional inter-
changes of nucleic acid segments may have occurred, and would have represented the dawn of genetic recombination, which is at the core of sex. When
the first cells established themselves, with high integrity, behind their plasma
membranes, they nevertheless would have been able to flirt by absorbing bits
of DNA or RNA cast into the surrounding environment by other selves, as
bacteria still do today, picking up plasmids (see below) that floated near in the
watery milieu. Despite these sorts of unconsummated molecular dalliances
with others' genetic essences, reproduction as a whole was *asexual,* taking the
form of simple cell division; in prokaryotes this process is known as binary
fission.

Each bacterial cell houses a single chromosome; its thread of DNA forms a
closed loop (additional, highly foreshortened circlets of DNA, the plasmids,
often accompany the main looped chromosome, with the ability to reversibly
insert into it). Shortly before the cell fissions into two daughter cells, enzymes
associated with the chromosome catalyze its replication. In an amazing rolling
and stitching series of chain reactions, the original chromosome is faithfully
copied, and two loops are pulled apart as the plasma membrane expands, and
then pinches off in the middle, leaving each of the progeny cells with identical
copies of the parent genome. The two new cells are thus clones. And this *asexual* activity may go on indefinitely, to produce astronomical numbers of cellular carbon copies.*

Thus before the original sexual revolution, all reproduction consisted of cellular divisions that replicated clones of progeny—daughter cells—virtually identical to the parents of previous generations. Then the prokaryotes evolved a
crude version of sex. Many bacteria species are now known to develop brief li-
aisons, a process called conjugation, during which individuals transfer DNA. The
individual cells that participate in such mating are not called male and female,
nor do they exhibit characteristics of sperm and egg. Instead, within their particular species, they differ as to whether or not they carry a particular segment
of DNA known as an F factor (F for fertility).[1] Suffice to say, regarding the con-
jugal relations of bacteria, that they rarely, if ever, proceed to full consummation—
that is, the transfer from one cell to another of the complete complement of
genes on a chromosome. Researchers have found that bacterial intercourse
is almost always frustrated by some sort of interference, physical, chemical, or
biological.

Bacterial sex appears crude, imprecise, and haphazard in creating new combinations of genes. Extremely proficient and climactic chromosomal intercourse
awaited the development of true sexuality in eukaryotes.

---

*Occasionally mutation, which is a sudden, random change in part of a DNA molecule,
throws down a wild card that can change the evolutionary prospects of an organism's descendants. A mutation changes a gene and is passed on thereafter as DNA replicates; a mutation may change a particular trait of an organism that is encoded in a gene.

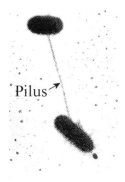

Pilus

Figure 5.1 Conjugation in bacteria. Cells of opposite mating types bind themselves to each other with a hollow cord called a sex pilus. The pilus forms an intercellular channel extending through the membranes of each of the mating individuals. Their cellular fluids are then in contact, and DNA may be partially transferred from one cell to its partner. The pilus is so fragile that it typically breaks shortly after gene transfer begins. *Source:* National Library of Medicine, Charles C. Brinton Jr. photo, public domain; image from plantphys.info/organismal/lechtml/bacteria.shtm.

Molecular Foreplay—Dances of Chromosomes

As we saw in the previous section, bacterial genetic recombination is typically incomplete. Through billions of years, nothing seems to have occurred to bring this feeble form of sex—a sort of coitus interruptus at the cellular level—to completion. And it was left to eukaryotic cells to take sex to its consummate form. This involved an ultimate precision at molecular and cellular levels, and then, in multicellular creatures, an explosion of wondrous and often bizarre expressions of structure and function. We pick up the story at a time when simple eukaryotic cell reproduction—each cell dividing on its own, maintaining its integrity with no sex in sight—had achieved its own precision, in the process called mitosis.

Mitosis is the way our bodies grow, and all eukaryotic cells (except for gametes in animals and spores in plants and some others) arise as clones, faithful asexual replicas. Moreover, these cloned eukaryote cells and organisms always constitute a particular stage of life in the cycle that oscillates between single- and double-chromosome complements (genetic endowment). Cells with a single set of chromosomes are called *haploid*, with the designation *n*. Those with a double set are *diploid*, or 2*n*, which, in animals, is the predominant life stage.

As beginning biology students learn, mitosis is a lot more elaborate than the rapid, looping chromosomal-replication-and-peeling-apart style of division of bacterial cells. Thus, eukaryotes, which at some early time had fragmented their DNA into numbers of individual threadlike chromosomes (compared with the singular, closed-looped forms of bacterial and archaeal cells), proceed in a very different manner to precisely replicate and distribute the copies of the genetic blueprint.

Somewhere in myth we might encounter a parable of a dance attended by self-cloning snakes that subsequently become amorous. On a certain evening, let us imagine, forty-six snakes appear in a small desert crater, ready to dance in a breeding ritual. Twenty-three are snakes identified with male fertility. They are of various sizes and banding patterns, and each of them is matched by a counterpart of female association, having a highly similar proportion and pattern.

Figure 5.2 Intimacy with snakes. The Snake Goddess of Knossos in ancient Crete would seem to represent an excellent patroness of sexual symbolism from the molecular to organismic levels. *Credit:* Drawing by Susan Culliney, inspired by a classic sculpture.

But as they first assemble, the snakes remain aloof; these individuals do not join physically as they now come together. And initially they appear as extremely long, skinny forms, but, as if in some backcountry of Alice's Wonderland, the snakes get shorter and stouter as we watch, and they *replicate* themselves, each becoming paired with an identical twin.

Moreover, this snake dance is self-organizing, without intercession by a Cretan goddess or Hopi shaman. As the music begins from a hidden source (flutes and drums seem appropriate), the twinned (replicated) snakes are side by side, held together at a particular site by a sort of molecular ribbon tied with a slip-knot; they all slither in their pairs around the crater.

The evening seems destined to be short, for these snakes know only two dances. The first is called *mitosis*. The music picks up in tempo, and all the snakes move toward the center of the crater, forming a line, still without any of the sinuous doublets touching. For a very short time, each doubled snake, female- or male-derived, is merely dancing with itself—a bit of writhing all along the line—and the slipknot holding each pair together pulls apart. Replicas separate evenly and, on each side of the line, forty-six single snakes now move away from each other toward opposite sides of the crater. It's over very quickly—until all the snakes again may replicate themselves and, in time, gather for the next mitosis dance.

The second kind of snake dance—*meiosis*—has a more complex choreography, for the snakes are feeling sexier. At the start of meiosis, after they have replicated identical copies of themselves, the doubled snakes experience a strong attraction, each to its opposite counterpart among the twinned pairs—the "male" tandem snakes (twin replicas tethered just as when they started out in mitosis) are attracted to join their particular *homologous* "female" partners in the center of the crater. Now, forming multiple Velcro-like protein attachments, the matched pairs bind precisely along their lengths—male- and female-heritage patterns (DNA segments) aligning perfectly. Clinging tightly together, they twist and writhe and swing around each other like square dancers for a while. Then their rotation stops, and they are lined up once more. The dancing ground is much less crowded now, with only twenty-three combined snake partners, instead of forty-six independently assorted double snakes. But now each of the twenty-three clusters is a tetrad, four snakes—two male replicates and two female—bound into intimacy.

In this dance, after the linked-in "male" and "female" snake doublets have ceased to rotate, they end up randomly polarized across the line. It's possible that all the males face toward the same side and the females face toward the other, but with twenty-three possibilities for either-way arrangements arising by chance from the snaky shake-and-shuffle along the line, the division about to ensue will yield one of an enormous number of different possible groupings of snaky banding patterns. Remember that each of the matched tandem-snake partners has unique features (typically small differences in the banding patterns that represent alternative forms of particular genes) related to their paternal versus maternal origin, and each has a fifty-fifty chance that the male or female partner-pair will face toward a given side of the line.

But the dance isn't over yet. The dancers now engage in an astonishing, kinky, further bout of intimacy. At up to several points along their lengths, where the Velcro-like proteins have attached them together, the twinned male and female snakes of each tetrad simultaneously break and exchange segments, becoming hybrids, crossing over their segments, donating and receiving them, and healing the breaks in exact register that reflects the matching patterns of the homologous tetrads. They have now merged their male heritage with the female in myriad combinations, and, as emergently mingled snakes, they peel apart and separate, still constituted as doublets. This time the original slipknots (unlike the Velcro-like attachments) have not parted.

Twenty-three hybridized snake-pairs—no longer the clones that started the dance—now migrate (meiosis has been called a *reduction division*) to opposite sides of the crater, but meiosis is not over yet. The performance concludes with one more round for each of the slithery clusters of twenty-three doublets. On opposite sides of the crater, each group lines up independently, as in mitosis. The trends of their lines are at right angles to the former linear array of the tetrads. Now the slipknots finally slip, and single snakes are freed and separate diametrically from each other, forming four clusters of snakes, each consisting of twenty-three single, harlequin hybrids of their former male and female derivations.*

---

*The analogy of dancing snakes serves to illustrate the precise distribution of eukaryotic cellular genomes and, in meiosis, a profound genetic shuffling. Structurally, both mitosis and meiosis are more complex than sketched in this crude analogy. For example, chromosomes, as they separate after their alignment (and, in meiosis, crossing over and exchanging segments), do not "migrate," or move by themselves. They are carried by kinesin (motor protein) molecules that "walk" along microtubule tracks, which themselves are arranged in a precise spindle pattern.

Aphrodite's Arithmetic

In the world of eukaryotic cell biology, the final clusters of harlequin "snakes" are the unit genetic complements of the haploid gametes. At the end of meiosis, the snakes are single chromosomes (or final versions of chromatids, as singlets are also known), the unit DNA molecules containing particular strings of genes representing the basic genetic endowment of each species. Through the partnering and recombination of male- and female-derived chromosomes in the dance of meiosis, haploid gametes end up with unique combinations of the genes of the original diploid organisms in each generation.

The power of meiosis to create genetic variability comes out of the deceptive power of two: eukaryotic cells commonly carry a doubled (diploid) complement of the genes of their species. But at a particular stage during the life cycle, meiosis precisely trims the genetic cargo of certain cells (gametes or spores) to the singular (haploid) DNA blueprint in a way that involves a profound shuffling of the deck of each individual's diploid complement. Then, culminating sex, in the form of mating and fusion of male and female gametes, establishes new diploid individuals in the next generation. They are never clones, but carry extensively recombined genes of their species, every individual different, even in a single family of offspring. And this shuffling continues through the generations.

In human gonads, out of the meiotic dance of the maternal and paternal chromosomes (twenty-three that came from the sperm and twenty-three from the egg that formed each of us) arises a huge number of genetic combinations in gametes. When human chromosomes arrange themselves in associated clusters prior to the first round of division in meiosis, the directional polarization (random male-female sidedness) of each tetrad relative to all the others presages up to 2^{23} possible combinations of chromosomes in the daughter cells after the division occurs. This so-called independent assortment of the tetrads results in close to 8.4 million different possible gametes in each individual. That goes for both sperms and eggs.

If nothing else were to happen beyond independent assortment in gamete formation, when sperm and egg combine after sexual intercourse to form a new human embryo, then $2^{23} \times 2^{23}$ parental chromosome combinations in the diploid complement of forty-six are possible in any given offspring. Thus, minimally, each of us represents one of about seventy trillion possible genetic beings.

But such calculations leave out the phenomenon of crossing over (which, in our parable above, involved the segment-swapping harlequin snakes) of the partner chromatids—with precise exchanges of DNA segments randomly occurring along the length of the molecular threads. This has the effect of vastly magnifying the genetic recombination in gametes beyond the simple shuffling of independent assortment. Many eukaryotic life-forms have fewer than twenty-three pairs of chromosomes, and some have considerably more. In all cases, the power of two plus crossing over of segments assures that a huge range of diversity accrues to any given species.[2]

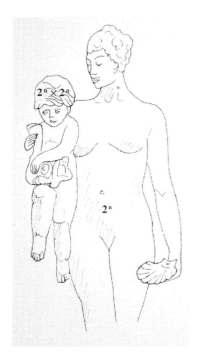

Figure 5.3 Aphrodite with Eros. Each of Aphrodite's haploid eggs is genetically distinct, representing one particular chromosomal combination among a minimum of 2^{23} possibilities. The child Eros represents the new diploid generation, combining twenty-three chromosomes from his mother with the same number from his father, and is endowed with a particular unique human genome emerging out of $2^{23} \times 2^{23}$ possible chromosome combinations, wherein parental genes typically have been further shuffled by crossing over. *Credit:* Drawn by Susan Culliney, inspired by a classic sculpture.

The Way We Were—How Eukaryotic Went Erotic

The wiles of the ancestral sexy eukaryote emerged one day, while all around that initial seduction, others remained celibate—in mitosis maintaining their integrity, merely replicating copies of themselves at intervals. The primordial protistan courtesan had acquired the property of attractiveness. At the time, the itch for union or fusion with a mate would have been strictly chemical, and many versions of that mode of attraction persist today. A rudimentary invitation to intimacy might have been as simple as a new type of protein (produced through a mutation in DNA; see more on mutation below), embedded in the membrane of the cell, projecting into the sea—a nanoscopic lure that would draw another of the same species into physical contact.

As we described in chapter 4, membranes of cells are extraordinarily delicate coverings, like the "skins" of soap bubbles. In cells that meet and touch surfaces, membranes are sometimes capable of fusing, and then dissolving, where contact is happening. Cells, like soap bubbles meeting in the air, can then open each to the other.

The initial phase of interaction that led to sex was likely just such a simple hybrid formation: two cells of a particular type of protozoan-like organism opening up to each other on contact and exchanging copies of their chromosomes (or, perhaps more likely, as seen in examples such as *Paramecium* today, whole replicated nuclei in the aftermath of mitosis). This initial move toward cellular intimacy through fusion would have led to an immediate physiological boost in the new cell's vigor and productivity, owing to the doubled number of chromosomes. Extra copies of genes can translate into greater amounts of proteins, the products of genes. Moreover, slight variations of particular proteins commonly appear in such new diploid cells; this happens because the codes for protein assembly in genes of various partners may have been changed slightly through differential mutations.

Cells gaining an extra nucleus from mating partners are not yet engaging in true sex. However, their new robustness would likely have provided them with an advantage in their world vis-à-vis competing cells that maintained merely a haploid set of genes (chromosome number = n). As suggested in the sequence below, at intervals, probable events that led descendant eukaryotic creatures to participate in near-universal sexuality oscillated between familiar ways of being.

- Following the *intimacy* of cellular fusion of outer membranes, nuclear *integrity* remained. A cell with two, or perhaps even more, integral nuclei became a powerhouse of chemical productivity, rapid growth, and clonal reproduction.
- At some point in descendants of this ancestral sexual being, two nuclei fused; their membranes gave way to a new, deeper round of *nuclear intimacy*, but *chromosomal integrity* remained. These cells continued to replicate DNA as before, but now with the $2n$ chromosome number. However, at this stage, the threads of DNA remained aloof from one another and replicated as independent units, as these cells continued to divide faithfully by mitosis, periodically cloning themselves and expanding the population of these new diploid cells as before.
- The cellular stage was set for a new process, *meiosis* via *chromosomal intimacy*—integrity giving way at yet a deeper level. Once again some sort of attraction-inducing change came about to bring chromosomes, themselves, together in pairs, embracing each other with stunningly precise alignments of their counterpart genes. The coupling involves pairs of so-called homologous chromosomes in the diploid cells. Each homologous pair consists of partners of the same size that carry copies of the same genes, albeit often in different versions (owing to mutations in DNA) called *alleles* that extend as segments through the DNA. So a component of the initial "mating" of homologous pairs of chromosomes that occurs in meiosis may be simply a high-fidelity fractal geometric congruence.

These events, or some similar progression, led to new states of cellular being that biologists today take for granted. Yet what we see here should evoke another deep sense of wonder regarding the power of cooperation in the course of evolution, and in its spawning of emergent potential.

So What Is Sex Good For?

Generations of biologists since the early twentieth century have agreed on a general Darwinian principle regarding the main evolutionary benefit of sex—that is, the combined effect of meiosis and fertilization (fusion of gametes). Seen as a very powerful conveyor of fitness to a species (but not predictably to any given individual), the benefit is in the genetic shuffling that comes with sex. In each sexual generation, as new combinations of genes appear, they shake up, mingle, and create alternative phenotypic expressions of individuals carrying those shuffled genes. *Phenotype* refers to the physical makeup, physiology, and behaviors that are encoded by the genes of any organism. Phenotype (in classical Darwinian thinking) is the basis on which natural selection operates, focusing on individuals but shaping the adaptations of populations.

Thus sexual genetic shuffling, as opposed to asexual cloning, keeps a species in the evolutionary moment by hedging its bets against environmental change and other selective pressures in the future. Sex perpetually maintains ranges of adaptive options, embodied by means of varying gene combinations among individuals in a breeding population. Some individuals in each generation will always have the right stuff to weather nature's frequently shifting challenges to survival.

Alternating between diploid and haploid states during an organism's lifetime also results in dodging lethal mutations and minimizing deleterious ones. Mutations as changes in genes may cause significant changes in proteins, which are the most diverse cellular products of genes. This may involve something new expressed in an organism, or something lacking that had been expressed or catalyzed by the original protein. In the first case, a new trait, such as sickle-cell hemoglobin in humans, may be carried but masked in a diploid genome. When a gene, as it manifests through its phenotypic expression, becomes defective,[3] its counterpart in a diploid genome can do its job, so to speak. Not all mutations are damaging; they hit DNA randomly, and without them life would be in ultimate stasis, for mutations are the agents of long-term, macroscopic change in biotic evolution and collectively become the feedstock of natural selection.

Thus, in its twin roles—continually refreshing genomes with new combinations of genes, all of which may eventually undergo mutation, and guarding diploid organisms against deleterious or defective phenotypic expressions—sex would have been favored in a mutation-prone world. Sacrificing of self—with its integrity of cloning—made sense, and sex may have evolved at a time of unusual mutagenic intensity, such as increased cosmic ray bombardment of the Earth. Nevertheless, one of the most profoundly emergent outcomes of the strange dance of the chromosomes in meiosis has been the flowering of a stunning diversity of life.

Sex Led to Many Splendored Things

The most flamboyant displays of sexuality in the biosphere are those of angiosperms, the flowering plants. These plants began their ascendance toward the end of the time of the dinosaurs, during which landscapes were largely dominated by conifers and ferns and a few others, some having impressive stature but consistently cloaking the plains and valleys and mountains of the world with shades of green. Even the reproductive structures on those ancient plants are primarily green, sometimes yellowing or browning with maturation. Among the first widespread angiosperms were magnolias; those growing today represent living fossils that have changed little in tens of millions of years.

Flowers are the sex organs of angiosperm plants. Their vast diversity of forms, colors, and scents has been created by mutation and natural selection largely in response to crucial intimate relationships with pollinators, especially insects. Not only honey bees and numerous solitary bee species, but many moths, beetles, flies, and others deliver pollen to precise locations where contained sperm gains access to waiting eggs. Besides insects, several other kinds of animals, notably various birds and bats, are known as pollinators of certain angiosperms. These plants have coevolved, each with its particular type of pollinator. For example, plants whose matings are assisted by moths typically have white or cream-colored flowers that open at dusk and emit powerful musky fragrances to lure their matchmakers in the night.

Striking coevolutionary congruence emerged in Hawai'i between nectar-feeding birds and certain plants. In one example, the flowers of particular lobelioid species evolved a curvature that exquisitely fits the beak of a native honeycreeper, the scarlet 'I'iwi, *Vestiaria coccinea*, as Herman T. Spieth vividly described in the 1960s.

> The nicety of fit between the bird's head and bill and the fleshy corolla of the flower is indeed striking. . . . The entire bill and fore part of the head are thrust deeply into the corolla, reminding one of a finger slipping into a well-fitting glove.[4]

This bird's efficiency of feeding in these peculiarly shaped flowers translates into highly effective pollen transfer for the plant. To be sure, not all flowering plants depend on animals' help to achieve pollination. Some, such as grasses, are wind-pollinated, with floral structures adapted accordingly, but the landscapes of the planet would be impoverished indeed without the wondrous floral displays that are attractors of other selves in the service of sex.

Among animals, sex organs themselves are not often eye-catching or colorful (female baboons are an exception) and poets may not extol their fragrances, which, though attractive to a potential mate, may range from pungent to extremely subtle. However, in a great variety of mammals, birds, fishes, and insects, sex has generated striking evolutionary differences in bodily forms and color patterns, as well as very diverse extra-gonadal chemical attractants, vocalizations, and dramatic quirks of behavior between males and females. Sexual dimorphism in animals ranges from barely detectable to extravagant. The differences that characterize males and females within a species are called secondary sex characteristics.

Sexual selection, a phenomenon that Darwin came to understand as "a struggle between individuals of one sex, generally the males, for the possession of the other sex," is the basis for the dimorphism. Darwin's reasoning was that the phenotypic particulars involved in the competition for a mating partner, as well as the lures responsible for triggering sexual attraction, arose in a manner parallel to natural selection. He imagined that over generations, such traits would be selectively honed for their effectiveness simply because, in Darwin's words, "The result is not death to the unsuccessful competitor, but few or no offspring."[5]

Thus the myriad manifestations of animal maleness and femaleness—bright plumage of birds or striking chromatophore (cellular-based pigmentation) patterns among fishes, displays of crests and antlers, songs and calls, chemical lures, elaborate courtship rituals—all of them unique and defining of particular species—have appeared in the service of sexual fulfillment, the mingling of the genes of a self with those of another.

Close-up, some of the humblest of creatures reveal the most astonishing sex lives to biologists. In the Hawaiian Islands, hundreds of species of native *Drosophila* flies (beyond the familiar kind that is attracted to overripe bananas in kitchens nearly everywhere) have evolved over millions of years. The many Hawaiian species are adapted variously to a great range of natural ecological settings among the islands, and rarely venture anywhere near bananas (which are not native to Hawai'i). Among the most intriguing aspects of the biology of these small flies, however, is that each species has diverged via sexual selection to embody unique patterns of courtship behavior. An entomologist's almost breathless description of the mating dance in one of the Hawaiian *Drosophila* species vividly describes the antics of the male, "which has evolved huge scent-dispersing brushes at the end of his tail which he curls over his head and shakes at his ladylove to overwhelm her with a shower of aphrodisiac perfume."[6] Without the genetic sacrifice and sharing of sex, the world would be an immeasurably duller place.

The Social Value of Sex

Viewed from a classic perspective in biology, the cooperation of male-female sexuality always leads back to the ethos of integrity, as embodied in each individual offspring. Survival of the individual, answering the challenges of natural selection to the point of reproducing and contributing to the heritage of the next generation, has been seen as the paramount outcome of life—what Darwin called "the struggle for existence." But perhaps a further subtle benefit of sex lies beneath this vision.

A key tenet of modern evolutionary theory is that all organisms are programmed to promote and defend their genes. The emphasis on integrity, hovering somewhere between genotype and phenotype, is clear. The neo-Darwinian theorist Richard Dawkins' concept of the selfish gene (see also chapter 6) crystallized some biologists' thinking that competition among genes is primal and tends to drive, by means of natural selection, the accumulation of traits that maximize certain sets of genes—those that Dawkins termed the best "replicators"—in a particular species. Such traits include rapid growth, adaptations to accumulate territory and food resources, and myriad means devoted to aggressive interaction—competition with other individuals of one's species.

Ironically, the selfish gene theory may help explain the evolution of altruistic behavior in social groups of closely related individuals. Cooperation, or even suicidal self-sacrifice of an individual in aid of the group that shares a significant proportion of its genes, becomes a selfish act at the level that favors the perpetuation of those genes—as distributed among surviving relatives. Neo-Darwinians have suggested this phenomenon evolved out of "kin selection" that is especially strong in social insects, such as honey bees and ants (see chapter 6). In these animals, competition and aggression among individuals in the group become displaced. Individuals are intimates (and siblings), but intimacy again gives way to integrity when it comes to different beehives and ant colonies that represent different genetic entities of such "superorganisms" (we return to these concepts in detail in chapter 6).

However, consider a problem in the long-term outlook for an aggressively competitive species. The African lion is a dramatic example. As the only species of cat to form social groups, wild lions have been subjects of extensive and exhaustive field studies by behavioral ecologists for over forty years. Scientists' shifting views of the basis for lions' sociality have considered efficiency of hunting on open landscapes as well as communal rearing of young through nutritionally lean times. Both are plausible as key selective forces that maintained the animals in prides of (mainly) closely related individuals. However, more recent studies[7] have strongly suggested that defense against other lions is the most powerful driver and shaper of their social organization as a species.

In lions, the problem stems in part from young adult males that commonly leave their pride (often driven off by older resident males). Footloose, the young lions roam together in small bands of brothers or cousins, and they begin to look for an unrelated pride that they can take by force, thus acquiring a stable hunting territory and a harem of females. When they take over a pride, the conquering males kill or drive off the resident (unrelated) males of the pride and then kill cubs that were sired by those defeated males. The new rulers then mate with the surviving females; sometimes one or two dies defending her cubs. However, most females acquiesce in the takeover. Even a stable pride within an established territory is often threatened along a "declared"* boundary with a neighboring pride. In this case, females are the usual targets and are attacked with deadly force by (largely) male members of the adjoining group. Access to more abundant prey may trigger this aggression into a neighboring territory, and the size of a given pride is clearly a major defensive asset on a savannah where up to several lion territories share common boundaries.

But this raises a question regarding the rationale for selfish genes, as held in highly competitive (aggressive) individuals, with respect to long-term evolution. Initially, as sex evolved, cloning would have made much more sense in the context of competition and maintaining the "most-selfish" genes in populations.

---

*Lions declare territorial boundaries by roaring, scent-marking with urine, and sometimes physical force.

However, genes that confer other adaptations useful for coping with climate change, exploiting new food resources, enhancing disease resistance, and so on are all potentially highly valuable.

So, if competition were primary (in the strongest Darwinian sense), cloning, not sex, is always going to preserve the highly aggressive, aggrandizing tendencies of organisms at the highest rate and for the long term. Thus perhaps sex, because it dilutes any extreme genetically derived tendency (and this happens quickly, within a few generations), is a "Gaian hedge" that saves sexual species from indefinite dominance by the over-aggressive, most competitive individuals. In the absence of meiosis and recombination, those individuals would tend to overwhelm species populations, monopolize resources, and simplify gene pools by causing extinction of many other genomic options.

The strange phenomenon of sexuality has been maintained with near universality in the evolution of eukaryotic life-forms, indicating a strong selective advantage. Its control of heritability represents part of the realization of Darwin's insight that he termed "descent with modification." Sex, in returning to a cooperative (or participatory) ethos, sets up a sort of Lockeian versus Hobbesian tension in the biosphere. Sex thus represents a *defending* mechanism of long-term biodiversity, versus cloning that would trend strongly toward competition. Although competitive traits never go away, their intensity and distribution are moderated, and sexuality powerfully lowers the curve of increasing aggressive tendency within any species. Sex as a key principle of biotic intimacy is a safety valve in nature against an ultimate, self-destructive integrity of aggressive replicators that would diminish life's complexity and turn evolutionary expression toward a brutish simplicity. In the long run, sex may protect against the threat of bottlenecking gene pools with populations envisioned by Tennyson—"red in tooth and claw"[8]—having impoverished endowment for survival in a world of change on many fronts.

Through long passages of evolution, many kinds of animals have explored and confirmed, through biological selection, sexual options deviating from straightforward male-female congress that results in the fertilization of gametes from a pair of parents. Behavioral cooperation of sexy selves extends through a spectrum of emergent partnerships: bizarre hermaphroditism in flatworms and snails, sex-changing fishes, homosexuality with many an unusual twist; sexual helpers and sneaks; threesomes that raise more young than a parental pair. Phenotypic expressions manifested in some of these sexual beings might have initially shocked, then delighted, Darwin. Nature has bestowed an exuberant sexual liberty on many of its creatures.[9]

In human society, on an emergent level, sex is antithetical to authoritarian control, especially by male-dominated organized religion. Sex is portrayed as evil, or sinful, by those who fear its psychically liberating sequelae. Male domination proceeds out of fear of loss of control vis-à-vis the other (women) and thus diminution of the power and aggrandizing social integrity of men. Then, too, those religious organizations and other social systems that favor male dominance and repress women appear, more often than others, to manifest sexual aggression toward women and children, psychic disjunction of self and other, and violence between men and women.

The psychic disjunction of self and other has often appeared in the traditional friction between heterosexuality and homosexuality. However, in much of current society we find sexual freedom reemerging with increasing levels of acceptance for gay and lesbian lifestyles, except, of course, for the reactionary religious-conservative crusade against this freedom. In ancient Greece, the philosopher Plato speculated in his *Symposium*[10] that homosexual eros was a higher form of desire than its heterosexual counterpart. In Plato's thinking, our desire for immortality is what drives us to reproduce (hence to heterosexual expression). The great philosopher would not have clearly understood that any particular eukaryotic biological legacy declines exponentially toward the vanishing point in subsequent generations, but Plato also considered a different sort of immortality. Homosexual love, for the Greeks, was commonly between an older and younger man (Plato did not speak much about lesbianism). The older man was seen as more of a cultural-inspirational guide and intellectual mentor to the younger man. The result of this relationship was progeny of a different and a higher sort, embodying memes instead of genes, as aesthetic and emotional sensibilities and knowledge of the world were passed along. Those sensibilities might well have greater influence going forward than the mere transmission of half of one's genes.

Far from Greece, in Hawaiian society, prior to contact with New England missionaries, male homosexuality seems to have been considered normal and enthusiastically embraced by both commoners and the nobility. More is known about the latter, in which typically young men were chosen by chiefs as personal retainers. Captain Cook and his officers during their ill-fated sojourn at Kealakekua Bay noted the openly affectionate associations between senior chiefs and their close companions, known as *aikane*. Although the English may not have been fully aware of the sexual context, they did comment on the apparent deep friendships of the same sex couples, and some of the young men acted as political intermediaries for their patrons.[11] An online search of historical and anthropological sources quickly reveals that, in many other places, widely divergent people such as Native Americans, pre-Islamic Indonesians, Australian aboriginal societies, and various native African cultures, prior to religion-driven colonial suppression, treated same-sex relations as normal. The spectrum of sexual practice in nonhuman primates, such as bonobos (see also chapter 6) suggests homosexuality evolved in prehuman lines of descent and naturally became part of human nature.

The benefits that emerge from free *intersexual* participation and contribution in the modern world include tolerance of diversity, tension reduction and conflict resolution among individuals and social groups, nurturing of affinity groups, and gains in trust and cooperation in many kinds of institutions, from families to corporations and other social groups, even including military organizations.

As we have followed the evolutionary path of complexity, our vastly diverse and intricate world has emerged with irrevocable force in waves of cooperativity that have transformed the physical planet we inhabit. The latest wave, that of human cultural expression, which includes sexuality, follows the same path. All the way from molecular interaction to cultural innovation, the initial step between self and other is *tolerance*. Next emerges a stage of *affinity, acceptance,* or *attraction,* and finally *appreciation,* in the sense of consolidation of gains or the enhancement of fitness as favored by natural or cultural selection. Against this progressive expression of the cooperative constant, reactionary resistance, with its source in hidebound integrity, cannot prevail for long in a universe that has embraced intimacy from the beginning.

Social Order in Nature
Between Conflict and Cooperation

> Any animal whatever, endowed with well-marked social instincts, the
> parental and filial affections being here included, would inevitably
> acquire a moral sense or conscience, as soon as its intellectual powers
> had become as well developed, or nearly as well developed, as in man.
> —Charles Darwin, *The Descent of Man, and Selection in Relation to Sex*

On every level, the universe is mutable and contingent, and those properties have woven themselves throughout time in the service of evolving patterns of matter and energy that we now try to measure and understand. Anyone reading this book is aware by now that a central theme is the interplay between cooperation and competition in universal evolution, and we have argued from the beginning that cooperation has been responsible for generating ascendant complexity.

We noted in chapter 4 that if there were "laws" of ecology, one of the foremost would be that life evolves to avoid conflict whenever and wherever possible. Of course, numerous exceptions occur among animals where direct conflict—for resources such as territory and food and, by males, for access to females—is observed. The greater the potential for a lethal outcome, however, the more likely that ritual posturing, aggressive display, and the like will evolve, in most cases, to resolve a dispute before actual fighting ensues. Many plants also compete aggressively, albeit slowly, for territory. Upward and spreading growth aboveground will tend to shade out competing individuals of the same or another species. Underground competition ensues by means of allelopathy: toxic secretions of some species' roots deter the roots of others in a forest or grassland. Mostly, this results in wider spacing among various species in a long-term, evolved community, but certain invasive species may temporarily take over much of a mountainside by aggressive growth and allelopathy. Although competition is universally acknowledged as a key principle at the heart of population and community ecology, and ecological patterns emerge that are attributed to competition, there is little evidence that competition proceeds very far in shaping complexity in nature. Instead, moving away from competition becomes favored by natural selection. Emergent pattern in such cases arises out of the *potential* for competition that, in most situations, seems to favor avoidance mechanisms that are open to natural selection and really responsible for observed variations of specialization, adaptive shifts, and new ecological roles of species in communities.

Avoidance of competition among different organisms in similar settings and with similar ecological needs might be construed as an indirect manifestation of the cooperative constant. As such, it supplements direct coevolutionary relationships between self and other that take the form of symbiosis known as *mutualism.* Furthermore the sequelae of competition-aversion among species having similar ecological niches, together with mutualistic symbioses between dissimilar species, appear to be important contributing factors responsible for Earth's biodiversity—distributed through many ecosystems filled with varied life-forms that have adapted to each other and to the physical conditions of their particular surroundings on the planet. The burgeoning of biodiversity, with occasional setbacks owing to catastrophic environmental impacts (some of them from outer space), has proceeded across about 3.7 billion years as an immeasurably branching tree of life.

In this chapter we look at patterns of coexistence and mutual support that appear at the most complex levels in the biology of animals, starting with a brief discussion of how evolution governs interrelationships between species (standard fare in introductory biology courses). Then we proceed to examine more intimate roles and interactions of individuals in populations of social animals, finishing with a close look at our nearest relatives among the great apes.

How do animal selves react to the *other?* Any one of three general modes of behavior will govern a particular encounter and lead to a specific action or posture. A self may behave with antagonism, in which case it will compete in some fashion with the other, or will mount an attack. In many situations, aggression stops short of actual combat and resolves itself by means of display activities, such as those of male iguanid lizards of a given species doing "push-ups" and flaring colorful patches of skin on their throats in front of each other. Such signals are threats, and nuanced perceptions of them in the brains of the potential combatants appear to signal the likelihood of winning or losing in a fight.

Males of some mantis shrimps, one of the fiercest kinds of small animals, often engage in ritual combat as they compete for territory on a coral reef. The various species that stage tests of strength to establish dominance have a pair of "raptorial" appendages with densely calcified knobs that deliver karate-style blows. The largest individuals can strike with a force approaching that of a .22 caliber bullet, and these predatory crustaceans kill crabs and fracture shells of snails and clams with this equipment. In addition, males of some of the larger species possess an armored, flaring tail, or telson, often with patterns (unique within a given species) of brightly contrasting patches and rings that are sometimes reminiscent of archery targets. While confrontations are often not set aside by mere display, fighting in these kinds of mantis shrimps, known as smashers, commonly devolves into an endurance match. Each male proceeds in turn to present its tail as target to the other, who takes his best shot at it.

Sometimes several rounds ensue until one of them has had enough and flees the contest.[1] The US military has even designed experimental personal, shock-absorbing armor based on mantis shrimp telsons.[2]

In the second general instance among animals, a self commonly exhibits a neutral reaction to another if the other is not a predator to be avoided. In nearly all cases, such tolerance between species also depends on the other not having a highly similar way of life in terms of habitat preference and food and other resource requirements (an organism's way of life, in its totality, is called its ecological niche). Herbivorous damselfish such as *Stegastes marginatus* on a coral reef will defend individual feeding territories that contain preferred algae that the fish graze. However, while they aggressively drive away any passing herbivores of their own or other species, they ignore similar-sized fishes that are carnivorous or scavengers.[3]

Tolerance sometimes goes as far as a commensal symbiotic relationship of two very different animals in close, continuous association, such as a pea crab with a large "innkeeper" (echiuran) worm.[4] In this case, the innkeeper maintains a burrow that shelters its crab guest, who appears to pay no sort of rent, but is innocuous to its host. Prairie dogs play a similar ecological role as host to an odd assortment of tenants, including snakes, in North America.

Then, beyond tolerance, there is active cooperation between dissimilar species. Many examples of mutualistic symbiosis, such as the algal-coral relationship that enables the construction of coral reefs (see chapter 4), have been discovered by naturalists and ecologists. Another marine example of mutual cooperation, famous in the annals of symbiosis, is the behavioral compatibility between small tropical cleaner-fishes (wrasses in the genus *Labroides*) and a range of larger reef fishes, including predators such as jacks and moray eels. The little cleaners occupy specific locations on a reef, where they pick parasites from the skin of their hosts, even entering hosts' mouths, cleaning their teeth and "cheeks," and exiting through the flaps of their gill covers. Cleaners have never been seen to be eaten. And in one study, hosts appeared to be protective of a cleaner, even when a threat to the host arrived in the form of a larger predator. This was interpreted as a form of biological *altruism* between dissimilar species.[5] We will look closely at altruism in nature, including human nature, through the rest of this chapter. In a few large-brained mammals, the evolutionary trend to altruism appears to have led to the emergence of empathy and sympathy, properties of mind that infuse the life of the fractal self.

Kin Selection: All in the Family

We now turn to a pattern of evolved social behavior that not only avoids competition among individuals but also extends beyond simple cooperation (mutualism, in which benefits are thought to accrue to all participants). We have to account for the *evolution of altruism:* a Darwinian enigma. Darwin viewed altruism in nature as a "special difficulty, which at first appeared to me insuperable, and actually fatal to my whole theory."[6] He then offered a vague notion that selection might favor self-sacrificing tendencies in individuals if benefits accrued to close relatives. In *The Descent of Man* (1871), Darwin briefly returned to this topic with a general notion regarding group selection (see below), but the evolutionary challenge did not begin to resolve in scientific terms until more than a century later.

Even today, some philosophers define altruism as strictly a human cultural trait—assisting another or others, at a cost to oneself, as a conscious act and even a premeditated form of moral behavior arising with motive and intent. As such, altruism is devoid of selfishness. At first glance, this moral altruism appears different from what biologists have observed in nonhuman animals.

In biological, or reproductive, altruism, participating individuals have been called actors and recipients. It is the nature of the exchange that recipients benefit, but actors incur a direct cost that detracts from their survival potential, or chance to reproduce (individual Darwinian fitness). Thus the question facing evolutionary theorists is, how might this behavioral pattern evolve? Any heritable tendency toward altruism should be selected against—ultimately a dead end. Actors that develop such tendencies sacrifice or risk significant portions of their individual fitness. Acts of altruism sometimes end up being fatally self-sacrificial.

Thus, in the theoretical framework of Darwinism, the evolution of altruism* was a notably hard sell until the 1960s, when studies of some of the most spectacular examples of biological altruism showed that behavioral patterns of certain social species might be linked to shared genes. Our understanding of the evolution of this form of behavior, which always involves some level of self-sacrifice, has been greatly enhanced by studies of insect societies, such as those of ants, certain wasps and bees, and termites. How and why altruism evolved in these examples was for several decades interpreted primarily in terms of closeness of relationship of the members in the tightly knit (eusocial) colonies of these animals. By definition, eusociality involves parents and offspring living in dense, permanent aggregations, with overlapping generations and special castes of individuals—at the simplest level, reproducers and workers. Most colony members do not reproduce themselves and are often sterile, but cooperate in rearing young, finding food, and defending the colony. Eusocial insects' colonies are numerically dominated by those workers, ceaselessly active individuals that construct, repair, and defend nests or hives, as well as forage and deliver resources to nurture the reproductive queen(s), her embryonic offspring, and a generally modest number of reproductive adult males.

---

*Altruistic behavior plausibly proceeds out of hormonal effects—the influence of neurotransmitters that stimulate emotion in brains of complex animals—or merely simple self-sacrificial instincts fixed by neurological circuits in the likes of ants and honeybees. Reinforcement of altruism in large-brained social animals may also involve cultural transmission, but its origins likely trace to neurological evolution—emerging out of side effects of physiological regulators and mating-inducing chemicals. Thus biologists can find support that even "moral altruism" stems from a genetic basis and is subject to selection.

The evolution of hymenopteran insects—ants, bees, wasps, and the like—has been notable for several branches that trended toward a eusocial extreme. In many *haplodiploid* species, colonies have only one queen at any given period, and the workers share up to three-quarters of their genes by common descent. Half of their genetic complement will be identical when they share the same *haploid* father (males develop from unfertilized eggs, thus possessing a single genetic complement—see chapter 5). In other words, all of his genes are shared by all of his offspring, who are all sisters and typically sterile. Maternal heritage of the worker-sisters derives from the diploid queen. Her eggs shuffle the two contributions that compose her own heritage, having split them into haploid complements via meiosis (chapter 5). Hence worker-individuals share all of their paternal and, on average, half of their maternal genetic endowment. This means that any worker's sacrifice that benefits the colony or otherwise boosts *colonial* fitness for overall ecological and reproductive success largely favors her own genes. In such cases, as the theory goes, genes for altruistic behavior will tend to be preserved by natural selection.

Even without the conditions of haplodiploidy and descent from a single queen, in social hymenopterans and termites, the colonies become superorganisms, and their individual members exist in somewhat the manner of cells in an integral animal, interconnecting and communicating through their instinctively self-organizing communal structure. As Bert Holldobler, Edward O. Wilson, and their colleagues have pointed out, paragons of complexity emerge from simple rules governing individuals, mediated largely by chemical communication. In these small animals, *group phenotypes* of wholly new forms and behaviors appear at the colony level: elaborate nests and hives, bridges and rafts fashioned by cooperating ants, carefully tended gardens of fungus, and striking military-like maneuvers in the service of colony defense and warfare.[7]

There are even eusocial mammals, the exceedingly strange blesmols, or colonial mole rats (family Bathyergidae—neither rats nor moles, but more closely related to guinea pigs and porcupines). Two species have been discovered; they live in underground colonies in widely separated dry regions of Africa. The better-known naked mole rats are, within each colony, all diploid offspring—both male and female—of a single queen. Only from one to a very few highly inbred males are reproductive. Colonies typically contain more males than females. Both sexes become workers that raise young and dig long tunnels in chain-gang fashion, then transition to defending the colony as they grow older. Perennial inbreeding results in a degree of genetic relationship—up to 80 percent—within a colony that can exceed that of nearly all other animal populations, including eusocial hymenopteran insects.[8]

The concept of how altruism could evolve out of closeness of relationship became known as kin selection, and the mechanism of its fixation in nature is called *inclusive fitness.* These ideas arose with a gene-centered view. Within evolutionary theory they were developed through the mid-1960s and early 1970s by population geneticists, notably William Hamilton and John Maynard Smith.[9] Their application in sociobiology subsequently accrued widely beyond social insects—to

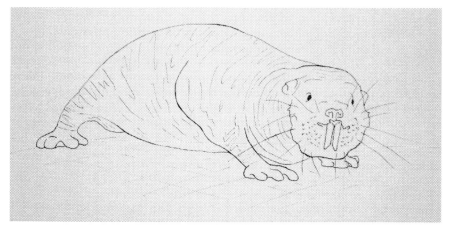

Figure 6.1 A naked mole rat, *Heterocephalus glaber*. This social mammal has evolved behavior and a reproductive pattern that parallel those of social insects. It cooperates closely with others in its kin group to excavate long, branching tunnels in semidesert habitat of Africa. Colonies feed on succulent roots and tubers encountered as they burrow. *Credit:* Drawn by Susan Culliney.

birds and mammals that maintain extended families exhibiting cooperation and clearly altruistic behavior as individual sacrifice directed within the most closely related populations. What Hamilton envisioned was that the related helpers or workers, whether they take risks such as facing predators while warning siblings or perform nurturing roles in the hive or nest or den, are enhancing the fitness of their own genetic identities, increasing survival and reproductive success of the very tendencies that bred them. Even if sentinels, workers, or soldiers sacrifice their lives for the colony, the evolutionary benefit regarding inclusive fitness of the kin group generally outweighs the mortal cost or genetic loss of the individual. However, Hamilton also recognized that ecological circumstances that favored aggregations of individuals, or the forming of colonies, would be important in setting the stage for evolving cooperation within the group.

Theoretical ramifications of kin selection subsumed in the theory of inclusive fitness began to infuse new support for a controversial earlier general hypothesis regarding natural selection operating on a higher level than the individual organism—namely that of local cohesively interbreeding groups within a species. Even Charles Darwin, in his sometimes neglected 1871 book, *The Descent of Man*, argued for group selection as a mechanism for the emergence of human cooperation.

When two tribes of primeval man, living in the same country, came into competition, if (other things being equal) the one tribe included a great number of courageous, sympathetic, and faithful members, who were always ready to warn each other of danger, to aid and defend each other, this tribe would succeed better and conquer the other.[10]

Nevertheless, until recently, most theorists since Darwin have consistently pointed to individuals as the key players in the game of natural selection—in which winning resolves itself in terms of differential reproduction. And any heritable tendency toward altruism that might arise would give way, as its carriers would be at a disadvantage to individuals that behaved selfishly. Those with the highest fitness in the context of their particular environment, and who profited from sacrificial behaviors of associates without returning the favors, would tend to leave more offspring. The altruists would lose ground in terms of lesser representation of their genes in future generations.

Then, amid the ferment of discussion on natural selection and altruism in the 1970s, Richard Dawkins' concept of the "selfish gene" gained many adherents in arguing that neither organisms nor groups of organisms, but rather genes themselves, competing as "replicators," are the logical units of selection, and the organisms built by genes are merely the means for their transmission, more or less successfully, into future generations (see also discussion in chapter 5). Dawkins, however, then evoked a kind of group selection operating among stable sets of genes, which endow their phenotypes with mutually adaptive characteristics. Such appropriately linked sets become units of selection. Examples can be drawn from animals with differing ecological constraints, such as carnivores and herbivores, in which quite different dental and digestive endowments, sensory requirements, and so forth are collectively conserved in DNA code.[11]

To many biologists, the argument has appeared to be whether the gene, the whole organism, or the cohesive breeding population represents the operative unit of selection. The issue might also be played out between the gene and the genome, with the latter encompassing both the individual level of the organism as well as the larger set of alleles (the alternative versions of specific genes, as discussed in chapter 5) constituting a group's genome within a given species. In the end, as first speculated in the 1960s, there almost certainly is not a simple answer, and all three levels of selection may operate in varying instances and to varying degrees in different species. Indeed, over the decades, evidence for multilevel selection has mounted. In the case of evolution of altruism, even in strongly eusocial species, inclusive fitness now is understood not to be the only explanatory mechanism (and perhaps not even the main mechanism).[12]

Reconsidering Cooperativity within Species

Through the last quarter of the twentieth century, inclusive fitness bloomed into dominance as the central and essential theoretical framework for sociobiology. Careers of many prominent academic thinkers in evolutionary biology were tied to what seemed to be the explanatory power of the concept, which some asserted approached that of natural selection itself. Inclusive fitness, defined as the degree to which an individual's personal genetic potential to produce adult offspring is enhanced or diminished by actions of his or her associates of the same species, was being touted as being predictive of the direction of evolution itself. By that, the theorists meant the increase or decrease in frequency, within future generations, of alleles that conferred social traits of a population. However, beginning in the late 1990s, inclusive fitness theory began to accrue serious doubts regarding its assertions of universality. Vague assumptions and unsupported conclusions were challenged regarding the theory's predictions of evolutionary trends and causes of behavioral changes in populations. Subsequently, problems with logical and mathematical rigor in interpretations of those changes also came to light. Results of some well-designed experimental studies on social interactions of real organisms falsified the theory, and more cogent explanations for such results have appeared amid great resistance on the part of conservative adherents.[13]

One of the central arguments regarding the weakness of inclusive fitness theory is that correlation of evidence is commonly presented as causation. Among examples given by Edward O. Wilson is one in which a particular allele or combination of them leads individuals to be attracted to certain members of their population that already possess a high level of personal fitness. Wilson refers to those that express this attraction trait as "hangers-on" and posits that their eager proximity to high-fitness individuals actually has little effect, if any, on the fitness of either category. But in the blind calculations of inclusive fitness theory, the association seems to reveal the hangers-on as cooperators, conferring enhanced fitness on those they are attracted to. Wilson concludes, "Of course this gets causality backward—the high fitness causes the interaction, not the other way around."[14]

With its logical failings becoming clear and predictive power all but lost in a wide range of real examples in nature, inclusive fitness theory may end up narrowly refocused on special cases and widely replaced with a variety of evolutionary explanations based on observations of details in the biology of particular species. Specific ecological attractors that promote close association and interdependence of species members, together with life history traits that assure cohesiveness in populations, may well be the primary drivers in the evolution of eusociality. Kin selection could then reinforce the emergence of altruism within local, inbred groups. But it could also weaken overall collective unity if a rival, secondarily inbred group arises within the original population. This often happens in colonies of some hymenoptera with multiple queens. Much controversy on these matters has arisen in the last decade. Some recent thinking has suggested that, ranging from simple extended families of animals to strongly eusocial colonies, evolution toward cooperation within a group that confers a selective advantage on the group may be driven nearly entirely by basic ecological circumstances regardless of the degree of genetic relatedness.[15]

A particular trigger could set a species on a pathway to a eusocial lifestyle, as discovered in the 1990s for the first time in crustaceans—in that case, certain small shrimps that colonize large tropical sponges. Snorkelers swimming over coral reefs hear these "snapping shrimps" making an almost continuous crackling sound. Among numerous kinds in the genus *Synalpheus* that compete for symbiotic living space within the labyrinthine interiors of a variety of shallow-water-dwelling marine sponges, eusociality has arisen in several species whose life history lacks a free-swimming larval stage. That appears to represent the trigger. Such stay-at-home species form large colonies that become strongly inbred. Many others whose young, as free larvae, drift for days to weeks in ocean currents are widely scattered by the time they are ready to colonize a sponge. They subsequently live mainly as isolated individuals that breed during random encounters with single partners of their species that are not close relatives. Research thus far shows that the colonial inbred kinds whose young develop within the colony form a unique subset of the genus that overall harbors at least thirty-five species in the Caribbean region. Taxonomic evidence suggests that the same developmental trigger opened independent pathways to eusociality in three evolutionary branches of the genus *Synalpheus*. In those emergent crustaceans, breeding has been confirmed to be reserved to a queen, and sterile workers help maintain and defend the colony inside the sponge. If Darwin were alive today, he would love this example of natural selection working through the complexities of ecology and life history with the confirmation of genetics and relatedness and emerging in altruism.[16]

Edward O. Wilson has recognized that among nonsocial hymenopterans, cooperative behavioral tendencies can be "spring loaded." He reports, among various examples, on certain solitary bees that typically live alone as unrelated individuals, which occasionally are forced to come together and then "they spontaneously divide up tasks such as foraging, tunneling, and nest guarding. Each bee is already programmed to perform these tasks on its own, switching among them as needed, but when a bee with a group encounters another bee that is already performing a task, it moves on to other tasks."[17] Further, it has been recognized for some time that, outside of eusociality, with its apparent synergistic combination of aggregative (colonial) ecology with inbreeding that may heighten inclusive fitness (with perhaps subsequent selection of altruistic traits), the increase and maintenance of cooperative traits of individuals in certain populations might be fostered in more generalized ecological settings such as localized nesting sites (e.g., the sponge-hosts of colonial synalphid shrimps), patchy feeding grounds, and so on. Those particular groups would then be subject to selection vis-à-vis other groups lacking altruistic tendencies. Weak altruism could strengthen over time once the balance had tipped toward some degree of self-sacrificing cooperation in a small ancestral population (see the next section). And in such cases, there needs to be a probability greater than chance that both actors and recipients in a population have a particular altruistic tendency for the trait to evolve. Many theorists already agree that fairly isolated populations, with little emigration of residents or immigration of outsiders, constitute the best incubators for sequestering and sustaining genes that foster cooperation and altruism.

Another possibility is that group selection could be sporadic, arising and fading away at intervals in various kinds of social animals, but not always maintaining strong coherence of selective advantage to build upon and sustain itself through evolutionary time. But there is no doubt today that evolved cooperation, altruism, and group enterprise, even when applied by tiny creatures

that at first appear insignificant, can mightily shape the biosphere. Those cooperators such as the species of ants with the most elaborate caste systems and
communication that have risen to peaks of eusociality since the Cretaceous
period became a new force on Earth. And as E. O. Wilson has noted in several of
his studies, they have been exceptional in exploring new evolutionary pathways
into vast ecological opportunities and capturing much of the planet's terrestrial
biological productivity.

Reciprocal Altruism: Evolutionary Gamesmanship

Beyond what some researchers still regard as the locked-in altruistic expressions
of kin selection, graded by the closeness of relationship in the likes of eusocial
insects and naked mole rats, are more fluid patterns of association in species that
have a clear capacity for recognizing individuals in their group of common associates. Intimacy and cooperation are strikingly evident in the ecology of many
such animals, including canids such as wolves, and with apparent deeper manifestations in elephants, dolphins, and primates. Various birds, too, develop intricate cooperative interactions, with striking emergent behavior involving, for
example, food sharing in ravens and mobbing of predators by geese.[18]

In 1971, Robert Trivers proposed a mechanism, *reciprocal altruism*, that would
favor the evolution of tendencies for even *unrelated* individuals of a species to
sequentially exchange mutually beneficial goods and/or services regardless of
closeness of genetic relationship.[19] The hypothesized reciprocity involved relative equivalence in the loss of fitness, or risk (with, on average, no immediate
mortal prospect), by the altruist and the gain realized by the recipient. Also,
Trivers noted that numerous opportunities for the give and take must be likely
during the lives of participants. An important corollary is that cheating individuals that habitually avoid reciprocating can be identified and excluded or
"punished" within the group. Individuals of species capable of evolving cooperative tendencies by reciprocal altruism need to be able to remember specific
encounters with others, rewarding and refusing to help, as appropriate. Thus, in
closely knit and stable populations of relatively large-brained, long-lived, social
animals with a prevalence of interindividual recognition there can emerge a
form of "trust" among individuals within groups. When this happens, just as
Darwin envisioned, such a group achieves a higher evolutionary fitness than one
lacking a significant trend toward cooperation.

Perhaps surprisingly, one of the better documented examples of resource
sharing attributable to reciprocal altruism is the vampire bat, *Desmodus rotun-
dus*. As noted in a landmark study by Gerald Wilkinson,[20] during daylight small
groups of bats, typically composed of up to a dozen females with their young at
various stages of immaturity, occupy roosts in hollow trees. There is a varying
degree of interchange of individuals among roosts in a general, regional population, so the degree of relatedness remains modest within a given roost. These
bats feed only on the blood of large mammals, primarily cattle, but often fail to
obtain a meal on a given night. Unsuccessful foragers, however, are commonly

fed by luckier roost-mates, which altruistically regurgitate substantial amounts of blood to their hungrier companions.

Wilkinson determined that if a bat does not obtain a blood-meal within three nights, it is likely to starve. The results of his studies—with both wild and captive bats—showed the significant prevalence of nonrandom partnering with regard to regurgitation events. He was able to demonstrate that both kin selection and reciprocal altruism occurred with respect to blood-sharing. In the latter case, observations with unrelated captive individuals showed that those deprived of food were fed selectively by others. The pattern of sharing was far from random; the blood donors nearly always regurgitated to those that had fed them in an earlier exchange.

The simple cost-to-benefit balance that theoretically favors the evolution of reciprocal altruism and stabilizes it in populations of individually discriminating social animals was anticipated decades earlier than Robert Trivers' breakthrough application of the concept in sociobiology. A quirky offshoot of mathematics called game theory, in the 1940s, began to analyze various strategies used by individuals playing card games, chess, and the like to determine effectiveness, or payoff probability, over many iterations of play. Later the methodology was applied to business practice and economics and even governmental functions and military operations, which involved varying degrees of competition and cooperation. Influential compendia and discussions of many applications of game theory, including sociobiology and evolution, have appeared in publications of Robert Axelrod since the 1980s.[21]

The particular game involving paired individuals that Trivers found to be a precise parallel regarding outcomes of cost and benefit, to represent his idea of reciprocal altruism, is called Prisoner's Dilemma. In evolution, the benefits and costs are calculated in terms of higher versus lower reproductive fitness; in Prisoner's Dilemma they are reward and punishment. The game involves two prisoners, who are kept in solitary confinement; they have been arrested for the same crime. Their subsequent penalties will depend on whether they both remain silent (mutual cooperation) or whether one "defects" by implicating the other in the crime. Neither individual knows what the other will do. The reward versus punishment structure of players' actions appears in the Prisoner's Dilemma diagram on the next page.

Obviously, if the game is played only once, the smart strategy is to defect and hope that the other remains silent. But the reward-to-punishment outcome changes when the game is repeated numerous times. Numerical values can be awarded to each player in each encounter. Cumulative points represent the reward-versus-punishment score as the game proceeds. This has been calculated for various strategies, such as the ultrasimple *always defect* and *be generous randomly,* as well as more complex algorithmic formulations for individuals' choices through the multiple iterations. But in long strings of replays, a strategy called *tit-for-tat* always ended up with the highest scores for both players. The rules for this strategy involve always cooperating (remaining silent) in the first instance. Thereafter, each player simply repeats the others' action of the previ-

Prisoner's Dilemma

		Player A	
		Silent	*Defects*
	Silent	Both get light sentences	A gets off, B gets long sentence
Player B	*Defects*	A gets long sentence, B gets off	Both get intermediate sentences

ous round. Thus such players will retaliate (punish cheating), but are ready to cooperate again immediately after a change of heart by the opponent. And, statistically analyzed, tit-for-tat's long-term benefits persist indefinitely. The calculations for long-term cost versus benefit in Prisoner's Dilemma are the same as those that estimate the relative genetic reward via numbers of future offspring accruing to individuals that reciprocate with cooperative action, as opposed to cheaters, in cohesive populations of social species.

Now, some biologists studying the ways that living systems function on perhaps every level from molecular to social are pointing to the phenomenon of *distributed information processing* as fostering the emergence of co-operation through group selection.[22] We should note here, too, that this is the organizational principle behind supercomputing and social networking on the Internet. Science fiction writers have long referred to distributed information processing as "group mind." However, this principle in earthly life operates all the way down to microbial mats (see chapter 4). It is also the basis of group behavior in social insects. And consider an advanced, individual animal's brain: decision-making may not be the action of what we consider an integrated mind as much as it is the output of a quorum of particular neurons of the cerebral cortex engaged in responding to a task at hand. Cooperation at an underlying level—below what we perceive as the level of action—is the prevailing process. And the outcome manifests itself as emergence. Seen in this light, biological group selection may be a hallmark of the cooperative constant that has been operating for eons—crossing over from the RNA world when the first biochemical replicators became associated with autocatalytic chain reactions and reached the threshold of life.

A few billion years later, the feedback between cooperators and cheaters in Prisoner's Dilemma can be seen to parallel dynamics in sociobiology involving vampire bats and many others. Cooperative interaction may vary widely in intensity among individuals and in strength in supporting the success of populations through time. Researchers have especially pointed to primates with strong mutual dependence such as baboons, bonobos, chimpanzees, and humans, whose cooperative social systems are likely to have a range of behaviors that evolved by reciprocal altruism.

In social behaviors of hominid primates (apes and humans), we begin to see the highest expressions of the cooperative constant extending into individual minds and group behaviors. These phenomena seem to appear initially out of underlying brain processes that trigger emotion. However, the capability for unprecedented individual influence in social groups, and far beyond, in the world at large originated on a seminal threshold late in primate evolution. In the next section we glimpse familiar tendencies that became embodied in the prototype or archetype of the fractal self. These tendencies would emerge and strengthen over several million years, ultimately expand from Africa[23] across the world, and enormously enlarge their potential in behavioral, social, and ecological contexts.

The Intimacy of Apes

Modern scientific studies of the social behavior of the great apes have led to insights that have begun to close what was long perceived as an evolutionary discontinuity—and to many philosophers, an unbridgeable chasm—between human beings and our closest relatives. In particular this research has elucidated spectacular new understandings of the minds of chimpanzees and bonobos. Primatologist Frans de Waal has led the way in making many of the most revealing discoveries in this field, formulating and testing hypotheses and evaluating evidence regarding capacities for empathy and sympathy, intuition and recognition of mental states of other individuals, and what de Waal terms "building blocks" of morality in our nonhuman cousins.

Based on over thirty years of investigation, largely at the Yerkes National Primate Research Center in Atlanta, Georgia, Frans de Waal's main thesis is that human morality has evolved out of rudimentary forms that are still traceable in apes, and perhaps with even deeper antecedents in species, such as capuchin monkeys, that split much earlier from our common ancestry. Thus human cooperative and moral modes of behavior appear to be intrinsic evolutionary outcomes, honed by natural selection. Primatologists argue that the best of what makes us human comes naturally and is not mystically and antithetically superimposed over an essentially selfish basis of being, such that we must go against our basic nature and genetic heritage to maintain our most civilized adaptations.

De Waal's wealth of research, in collaboration with dozens of his graduate students and often with other primatologists and behavioral scientists around the world, has led the way to our current understanding of the biological basis of the highest and best qualities of humanity. The citations we have made in this chapter distill only a small fraction of this work.

A small anecdote that De Waal relates by way of introducing the implications that have emerged from his many long-term scientific studies concerns a female bonobo (named Kuni) in a British zoo, who, after catching a starling that remained alive but stunned, was persuaded by a zookeeper to treat the bird with gentleness. Soon thereafter, the ape climbed the highest tree in her enclosure with the bird carefully held in her hand. With her legs and feet clinging precariously to the top branches, Kuni stretched the bird's wings widely in both hands and flung the comatose captive into the air. Although the bird, unable to fly, fell to the ground, Kuni later guarded its body against manipulation by another curious member of her enclave.

In his telling of the story of Kuni,[24] Frans de Waal implies that bonobo behavior includes strong empathic tendencies that are shared with humans. De Waal believes Kuni was not merely treating the bird as a plaything (with far greater creativity than does, say, a cat) but was actively trying to assist the bird, at some risk and expense of effort to herself. Kuni appears to have acted out of awareness of the bird's need for help in a specific function that might restore its vitality and wholeness. Empathy then is the starting point for a state of mind in a self that imagines, or anticipates, a need or desire in another individual; sometimes such anticipation triggers an appropriate cooperation that is congruent with a perception of the other's well-being, even in the case of a member of a different species. Apes clearly have the capacity for empathy at a complex level; captive chimps and bonobos exhibit empathy in their relations with each other and with humans in many striking examples, and a gorilla named Koko became admired around the world for her care of a pet kitten.[25]

De Waal suggests that empathy, as "inter individual linkage" is primary in our ancestry as a first step toward developing a moral sense. As realized in chimps and bonobos, empathy involves an intuitive awareness of mental states of others. Out of empathy may emerge what various behavioral researchers have called a "theory of mind," in which an individual's awareness prompts his or her taking another's perspective, and de Waal argues that this hallmark of intimacy in evolution proceeds directly out of emotion. "At the core of perspective-taking is emotional linkage between individuals—widespread in social mammals—upon which evolution . . . . builds ever more complex manifestations, including appraisal of another's knowledge and intentions."[26]

Frans de Waal proposes that the fundamental capacity of empathy arises out of a state of "emotional contagion" that is seen in a variety of mammals and even some birds such as parrots, geese, and ravens. At the simplest level, one individual "catches" and reflects the emotional state of another.[27] In his insistence on the origin of empathy proceeding from emotion, de Waal breaks with the general view of psychologists and mainly Western philosophers, which held, at least until very recently, that empathy from the beginning was strictly a cognitive trait in humans and largely depended on expressive language in its origin and development.

Modern observations of primate social interaction confirm that chimps and bonobos carry their intuitive perceptions of other's mental states to the more complex level of *sympathy* that often evokes *actions* going well beyond empathy. Sympathy has been defined as "an affective response that consists of feelings of sorrow or concern for a distressed or needy other" and is "believed to involve other-oriented altruistic motivation."[28] Defined in this way, sympathy follows empathy, and its emotional state stems not merely from a direct assumption by one individual of distressed feelings perceived in another, but proceeds to quite different feelings of concern on the part of the sympathizer. Moreover,

Figure 6.2 Two young bonobos embrace in the Lola ya Bonobo Sanctuary, Democratic Republic of Congo. *Credit:* Photo by Dr. Zanna Clay, used with permission.

sympathy may then elicit helpful action. Both empathy and sympathy appeared to be involved in the story of Kuni, and various primate-behavior researchers have now recorded many other examples involving consolation behavior, succor, and rescue efforts among bonobos and chimps, arguably the most "gifted" of our nonhuman relatives.

The Morality Question

One of the earliest evocations of "the Golden Rule" appears in the writings of Confucius, who considered positive reciprocal action among individuals as the keynote of fairness, and definitive in a moral life. When Confucius was asked by one of his students if "there was one expression that can be acted upon until the end of one's days," he got this reply: "There is *shu* 恕: do not impose on others what you yourself do not want."[29] Later, the Confucian philosopher Mencius shifted the master's proclamation to the affirmative: "Try your best to treat others as you would wish to be treated yourself, and you will find that this is the shortest way to benevolence."[30]

Mencius further emphasized what he saw as an *instinctive* core of goodness in human nature. One of his celebrated examples points to an occasion of moral anxiety recognizable by nearly everybody.

> The reason why I say all humans have hearts that are not unfeeling toward others is this. Suppose someone suddenly saw a child about to fall into a well: anyone in such a situation would have a feeling of alarm and compassion—not because one sought to get in good with the child's parents, not because

one wanted fame among one's neighbors and friends, and not because one would dislike the sound of the child's cries. From this we can see that if one is without the feeling of compassion, one is not human.[31]

We feel that Mencius would immediately empathize with a chimpanzee that reacts to another member of its social group in danger. A number of observations of chimps living in zoos, where they are confined to islands isolated by moats of deep water, have occasioned close parallels to the child-in-the-well scenario. As renowned primatologist Jane Goodall has noted, chimps are generally afraid of deep water; they lack the ability to swim, but they have been seen to demonstrate an altruistic response as strong as that of the most admirable of human beings. "Individuals have sometimes made heroic efforts to save companions from drowning—and were sometimes successful. One adult male, however, lost his life as he tried to rescue a small infant whose incompetent mother had allowed it to fall into the water."[32]

Despite mounting evidence that chimps and bonobos have reached a threshold of moral sensibility, many scholars have seemed unable to cross an evolutionary bridge that links humanity to our closest relatives among the apes. Frans de Waal addresses classic views (still widely held) of humans as the only beings having a capacity for morality that uniquely separates us from the rest of the animal kingdom. He dubs this way of thinking "veneer theory," and traces its origin to Thomas Huxley, Darwin's great defender in the nineteenth century. Huxley's otherwise staunch defense of evolution through natural selection seems to fall apart at the ultimate sticking point—human descent-with-modification of a moral and ethical nature. Instead, in a major address, "Evolution and Ethics," in 1893,[33] Huxley instructed his audience: "Let us understand, once and for all, that the ethical progress of society depends, not on imitating the cosmic process, . . . but in combating it. . . . The practice of that which is ethically best— what we call goodness or virtue– involves a course of conduct which, in all respects, is opposed to that which leads to success in the cosmic struggle for existence."

In a follow-up essay, the "Prolegomena,"[34] Huxley used the metaphor of a garden to represent human culture, with its need of constant cultivation to maintain a thin layer of order and civilization, which is surrounded on all sides by unruly nature. There is the strong implication in Huxley's remarks that he regarded human morality as unnatural, arising mysteriously as an overlay (veneer) on our evolved nature, which was competitive, aggressive, and inescapably selfish. This broke the link of evolution extending to human beings that Darwin had so meticulously established in *The Descent of Man*. Huxley's break with Darwin was, in de Waal's view, "that what makes us human could not be handled by evolutionary theory. We can become moral only by opposing our own nature."[35]

On this point, some modern philosophers and even some biologists who have considered the issue (among them, Richard Dawkins) remain strongly at odds with Darwin. When it comes to the question of evolution of moral behavior, Dawkins comes down on the side of Huxley and veneer theory.

My own feeling is that a human society based simply on the gene's law of universal ruthless selfishness would be a very nasty society in which to live. . . . Be warned that if you wish, as I do, to build a society in which individuals cooperate generously and unselfishly towards a common good, you can expect little help from biological nature. Let us try to *teach* generosity and altruism, because we are born selfish.[36]

Christine Korsgaard, among numerous contemporary social philosophers, has maintained that "there is some deep discontinuity between humans and the other animals." She concedes that apes—in the moment—exhibit a form of empathy that can proceed to altruism. This behavior arises with intention and translates to action, but nevertheless seems to Korsgaard solely a product of emotion. The significant behavioral gulf that exists between apes and humans lies in the terrain of conscious reflection, which extends the capacity for moral consideration and the impetus for action from a specific encounter or event to the general case, or ideal ethic, of an individual vis-à-vis society. It is Korsgaard's view that the self's ability to reflect on socially beneficial behavior for the long term "and the deeper level of intentional control that goes with it is probably unique to human beings. And it is in the proper use of this capacity—the ability to form and act on judgments of what we ought to do—that the essence of morality lies, not in altruism or the pursuit of the greater good."[37]

Of course humans very often fall short of this ultimate internalization of the Golden Rule, and we much more commonly demonstrate the impulsive-cooperative, sympathetic sorts of behaviors—those identified by primatologists as building blocks of morality in apes: mutual aid, self-sacrifice, fairness, conflict resolution, forgiveness and consolation, and more—in our socially constructive ways of treating others. Perhaps in most cultures this summation would indicate that most of the time we still behave more like apes than humans. There might be a tenable argument that *Homo sapiens* have merely just arrived (evolved as an ape) to stand on the threshold of a new basis for extending and deepening social interaction. And this capacity is emerging out of our dawning reflections of what we *ought* to be doing in the world. Darwin's visionary voyage into human evolution long ago visited these shores in considering the highest attainment of morality in a social species; in his attempt to understand moral emergence, the great naturalist declared, "It is summed up in that short but imperious word *ought*, so full of high significance."[38] Then, too, Darwin considered evolution in general as a work in progress.

So much of moral behavior arises from emotion, as noted by Mencius, and then is strengthened by conditioning—the feeling of satisfaction and well-being that comes from cooperation, from teamwork as opposed to conflict, from getting a sort of "high" out of our rendering of "goodness" in relationships with others. This often appears to leave cognitive (calculated) morality as a selfish principle, especially when organized religion is involved (one thinks of gaining heavenly credit through performing good works). Thus is it obvious that the Judeo-Christian-Islamic viewpoint of human nature is firmly embedded in veneer theory. We are bad by nature, as envisioned in the doctrine of original sin; hence the need for redemption by divine, or semidivine, intercession—requiring a messiah to magically rescue us from our natural immorality.

This pessimistic view of humanity is not shared by various Eastern philosophical and religious traditions, even in cultural niches that at various times and on some levels strongly emphasize competition. For example, in the codes of *bushido*, the anticipatory assumption of a samurai, upon encountering another (including a retainer of a rival *daimyo*), is that the other is an honorable man.[39] This trust, hovering between integrity and intimacy on the part of the interacting individuals, is foremost in relationships. The interactive ethos of medieval Japan was a realization of tit-for-tat strategy in sociobiology. Of course, in samurai culture betrayal was not uncommon, and the initial cooperative expectation could be breached, triggering conflict, but the initial stance is one that confers hope for our basic, inherited nature. If, from the start, the self's expectation is that the other is "good-natured" in the sense of Frans de Waal, when considering the social lives of bonobos—and by extension the hominid ancestry that gave rise to us all—we are seeing the formative footsteps of the fractal self.

Transcending Tribalism

Frans de Waal views the evolution of altruism in primates, with its highest expression in morality, as a nested set of behaviors. He likens this evolved behavioral hierarchy to the classic hollow wooden matryoshka dolls that became popular in nineteenth-century Russia. The dolls fit one inside another and coexist snugly like the progressively more highly evolved levels of "moral" behavior that begin with stimulus-response instincts and end with conscious decisions. On the inside, at the center of de Waal's "Russian doll" model,[40] is emotional contagion triggering reflexive response—"aping" of another's joy, fear, unhappiness and other states. This stage is merely intuitive, and a responsive self subconsciously reacts to body language of the other—for example, facial expressions, blushing, pupil dilation, and the like. Mirror neurons, first discovered in macaque monkeys in the early 1990s, may be engaged in these responses at a reflexive level.[41] Earlier in this chapter we noted that a wide range of birds and mammals, including humans, express such behavior in varying ecological contexts. This core complex of intimate emotional connection then, in humans, may rise to the next level: cognitive empathy—an active seeking to understand the other's mental state—and ultimately to "attribution," which is the most complex state of assumption of another's mentality of the moment, modeling it within one's own consciousness. In this section and through the rest of this book we will see that the dawning of the fractal self in the human condition extends the Russian doll to a new outermost entity—namely, a prosocial awareness, pregnant with expansive conscious potential of human beings to empathize across scale through our affinitive worlds of culture and nature. However, the process is full of setbacks, in all societies and on many scales.

The evolutionary biologist David Sloan Wilson, known for his work on animal altruism and for revitalizing group (or multilevel) selection theory, has recently turned his attention to patterns of urban community cooperation. Exploring what he sees as a Darwinian model for guiding progressive emergence of human generosity, dubbed *prosociality,* led Wilson to study the state of cooperative consciousness in his home town, Binghamton, New York (Wilson is a professor at SUNY, Binghamton). In his recent book, *The Neighborhood Project,*[42] Wilson and his students mapped and surveyed neighborhoods across the city to rank levels of prosocial attitude. In Binghamton's schools, they used questionnaires for children to gauge their altruistic tendencies as well as perceptions of nurturing engendered within their own families and in the general support of neighbors. Even the distribution and intensity of holiday lighting displays at Halloween and Christmas in the various neighborhoods provided data for plotting contours of "social capital," a term from economics that in this study meant solidarity and cooperativity. If the data had been merely random around the city, the neighborhoods would all have merged in a flat landscape, but Wilson reported the contours resembled a "map of the Himalayas." The upshot of *The Neighborhood Project* is the hope that social capital will spread in the city by managed projects developed and funded by government and private civic organizations. The suggested Darwinian mechanism is friendly competition among neighborhoods to design and build playgrounds, propose new programs for children and seniors, and so on.

Critics of D. S. Wilson's social prescription for urban ecosystems to "evolve" away from problems such as blight, crime, and other community ills and toward a wider cooperative ethos suggest his group selection model is flawed. Factors far outside of local control—housing markets, private disinvestment and loss of employment, boom and bust, in some cases with catastrophic economic consequences—more nearly resemble the vicissitudes encountered in the ecological, not evolutionary, process of succession (see chapter 4). During succession, setbacks and, more rarely, catastrophes occur from time to time, and natural communities within their ecosystems gradually recover and rebuild complexity in the web of interaction among their species. Enormously important in the process are catalytic roles played by symbionts, such as lichens that build organic soil and nitrogen-fixing microbes that enrich that soil. Cities as well as nature's communities and ecosystems become more complex from earlier, simpler assemblages. They recover from setbacks and may shift direction, as is the case with some "rust-belt" cities in the United States now attempting to move away from collapsed economies based on heavy industry toward new enterprises, some of them with hopeful applications in a greener world. But we think that a city's shifting socioeconomic landscape through time develops largely by self-organization, and the turning and tipping points that accelerate and may enrich urban succession are often mediated by the kinds of people we call fractal selves. The next part of this book will explore their potential as catalysts of positive change in the world, within and far beyond our cities.

It is both a great strength and a weakness of humanity that our evolution has retained the roots of emotional empathy common to other social animals even as conscious and considered moral behavior emerged in our psyche as a species. Reflexive action that expresses Confucius' Golden Rule has the same basis as biophilia in general. Such behaviors commonly do not reach a state of deliberation in one's mind. As Edward O. Wilson and others have pointed out, they are triggered and acted upon out of deeply ingrained emotion.[43] Frequently, it seems, emotion leads in our treatment of other selves; then a cascade may or may not ensue through the more rarefied layers of the Russian doll rising into consciousness and toward the highest expressions of the fractal self.

With the attainment of that pinnacle potential of moral behavior, in the present moment of the world everything has changed. This represents an emergence of enormous portent, ranking perhaps with milestones such as the chlorophyll molecule and eukaryotic life. However, we are a work in progress; human minds still dwell much of the time deep in the core of the Russian doll, confined in highly parochial modes of behavior that proceed from emotional contagion. To complete our emergence as *Homo sapiens*, we need to cross those ravines that lie beyond our immediate kin groups and tribal affiliations. Our species has come to dominate virtually every ecosystem on the planet, once as near a paradise as we might imagine in our universe. We inhabit a biosphere that owes so much to intimacy, but whose fate is now bound to humanity's uncertain balance between cooperation and competition. As a prominent environmentalist put it recently, "Nature before . . . us is dead. We are in a radically new moral position, because we are at the controls."[44] To this a natural philosopher and Darwinist, coming full circle, might add, "And we *ourselves* require a sustaining biosphere. Without it *we* are dead." Our biological instincts, cultural tendencies, and technological metabolisms can either nurture or destroy living systems that have enabled our being. As described in the next section of our book, the way of the fractal self points out a hopeful path leading to resolution of our conflicts with our wondrously evolved environment and ourselves.

Emergence of the Fractal Self

Self within World

When I look for my existence I do not look for it in myself.

—Antonio Porchia, *Voices*

When may we begin to *"naturalize"* humanity in terms of a pure,
newly discovered, newly redeemed nature!

Nietzsche, *The Gay Science*

T hrough eons of evolutionary time and across scale from atoms to cells
to organisms to species populating ecosystems and, ultimately, to
evolving social systems of complex animals, the relationship between
self and other oscillated unconsciously. Cooperation and competition played their roles in an increasingly complex world, and hominin brains
began to focus on the meaning of self, of its origins, and how it related to the
surrounding systems of nature and to other individual human beings. Recall
that in chapter 2, seeking common ground in ancient creation myths of disparate peoples, we found that our formative thoughts and myths strongly
embraced holistic views of the origin and development of the world and cosmos;
they portrayed self-organization and emergence as immanent natural processes;
and they consistently featured humanity as naturally inherent in the evolution
of systems by which complexity becomes manifest. Through myth, our ancestors envisioned cooperative and catalytic action that emerged in new, dynamic
pathways of cultural evolution, bringing humanity into a potential, participatory,
role in the unfolding of the universe. We became sentient butterflies. And
these remarkably similar strata of ancient holistic thought, across widely separated cultures, seem to have remained at the surface for the longest time in
China.

In this chapter, we introduce the foundations of the fractal self in relation to
the Chinese notion of personal development and enhancement of adeptness in
the world and mutualism with the other. This seeking, described in the codified
system of Daoism, is a pathway that may progress to the highest level of achievement of such a self: that which defines a *sage*. A sage, in our view, is a fractal self
that achieves a peak of intimacy and constructive interaction with the world. We
will discuss the nature of sagely status and sagely behavior in chapter 8. Here we
detail the development of the fractal self, emerging beyond the core embodiments

of empathy, sympathy, and rudimentary morality observed in apes (chapter 6). The self, for the early Chinese, as we will see, was always a being that was embedded in the world and dynamic flow of forces. This self was defined in intimate terms as adaptable and adept, seeking to be a microcosmic contributor to some holistic macrocosm. In this chapter, Daoism guides our thinking on how the fractal self engages with the world and resonates with our discussions of the self in Buddhism and Pre-Socratic thinking in the West.

Roots of Chinese Worldly Philosophy

The Chinese were for the most part devoid of the kinds of creation myths that proliferated in the West. Although Shangdi (上帝) (literally, a "god on high" as well as high ancestor) was a patriarchal god, he was not related to cosmogonic beginnings and was one god among many of the ancestral and nature spirits. In other words, Shangdi was neither a creator god nor the dominant force in the universe; although remote, he was not a counterpart to Zeus, the primal ancestor of the Christian and Islamic God, who managed the universe from afar and cast events in the world with a nod of his head.

When the Shang people were eventually overthrown by the Zhou sometime between 1023 BCE and 1123 BCE, Shangdi began taking on more naturalistic characteristics. Philip J. Ivanhoe and Bryan W. Van Norden note that this development was "a preference for accounts of actions and events in terms of systematic, natural phenomena rather than spiritual power."[1] Both early and later Chinese religious thinkers would always reveal a "this-worldly" bearing in their attempt to explain the complexity of the world and universe in which they found themselves.

Eventually Zhou "theology" tended to collapse earlier Shang conceptions of Shangdi with their own "supreme" deity Tian (Heaven or Sky). Tian is often used as an abbreviation for *tiandi*, heaven and earth, showing the inclusivity and interrelatedness of the universe writ large as the participatory home of *Homo sapiens*. And on the microcosmic level of the early Chinese community, the worship of dead ancestors provided a certain godliness and sacredness to the human dimension as well as a continuum of the world for spiritual and mental companionship. Such a dimension led to the reification of living humans and their connectivity to both the social and natural systems in which they were embedded; in a profound sense, humans became gods themselves. As "gods," sacrifices to the ancestors became a practice, a powerful way of sanctifying relations between the dead and living, and reinforced the social status among the living within the family and beyond into the community. By this time, the greater waves of formative philosophical tradition in China were already metaphorically resonant with scientific theories of holism that would emerge only after many centuries. The tradition would provide a powerful counterpoint to the anthropocentric, deity-directed view of the universe and the subsequent byproducts of that view—primarily the dominant Western one that sanctions a higher reality, a transcendent god, and an immortal soul.

A clear reverence for nature also pervaded early Chinese thinking and accords with our understanding that, among all major philosophical traditions, Daoism perhaps most clearly recognizes and embraces self-organization in an evolutionary view of nature. Thus did holism, now embodied in modern chaos and complexity theories, appear at the center of the Daoist portion of the Chinese worldview. Daoist thinkers understood how the systems of nature flourish and diminish and how patterns emerge from the various interactive components of the natural world. In mid-twentieth century, Benoit Mandelbrot (see the introduction) confirmed, with elegant mathematical simplicity, that intricate, repetitive patterns appeared in the universe over a vast range of scale, and that nature created structure in a continuum of dimensionality.

Mandelbrot's insights are reminiscent of the approach taken by the ancient formulators of the *Yijing*, the Chinese classic *Book of Changes,* who seem to be the first to arrive at an understanding of interlocking patterns of the human and natural worlds. Their insight was to imagine "a system of coordinates, a tabulation framework, a stratified matrix in which everything had its position, connected by the 'proper channels' with everything else."[2] Chinese philosophers during the second and third centuries would also maintain that the seamless dimensionality in nature is the definitive characteristic of Dao—the way the world is formed and the way it behaves: "Way-making (*dao*) is the flowing together of all things (*wanwu*)," and "It is inherent in things that they are ties to each other, that one kind calls up another."[3] Dao (道), or "way," is in many ways just life itself, the flowing of life, or even the changing world itself. The "flowing together of all things" in the quote above is *wanwu,* the totality of all that is happening in the world. There is nothing mystical about Dao for Chinese philosophers—no God behind the scenes nor nonreducible substance behind the myriad manifestations of the diversity in the world. Dao, as we will see, is the

Figure 7.1 In a section of the 1244 CE *Nine Dragons Scroll* by Chen Rong, the seamless flow of Dao, suggesting emergence amid turbulence across the scales of the world is skillfully represented in monochrome ink on paper with elusive touches of red. *Credit:* Museum of Fine Arts Boston.

unfolding of all that is and is best understood as a gerund—"dao-ing" or "way-making." We will return more specifically to *wanwu* later in this chapter.

The emphasis of the Daoist view is on the natural qualities of Dao that arise from the principle of self-organization and the synergy of cooperation. Dao happens, but its happenings move in subtle ways that appear to be without purpose. Daoism recognizes an evolutionary process that generates pattern (order from chaos) within a system, or part of the world—a process that develops as if there were a ghost in the machine, a gestational force arising from the interactions of the system's integrated parts. This emergent tendency of nature to build from within an integrating structure that directs its entire system is Dao, the way of universal evolution: "The Tao [Dao] constitutes the regulating structure of nature."[4] In keeping with chaos theory, we could say Dao encodes the rules that govern the structure of nature as it operates through self-regulation and evolves through progressive self-organization.

Wuwei as the Cooperative Constant

Throughout this book, we have maintained that nature, meaning the unfolding of the universe, has proceeded to build pattern and complexity out of a tendency to cooperate among entities across scale. The Daoist concept *wuwei* (無為), which means literally nonaction, but carries connotations of spontaneous, natural, and nonassertive or nondirected action, emphasizes the orderly becoming of the world; it reflects the natural, unconscious, and nondirected action within an anti-teleological process model of the world to which Laozi, the mythical author of the *Daodejing*, or the *Classic of the Power of the Way* (also known as the *Laozi*), refers. The flow of the world (proceeding out of its seamless structure) is Dao in action, or "way-making."[5] The shaping of Dao then is understood as *wuwei*: in its self-organizing fashion, Dao "really does things non-coercively, yet everything gets done,"[6] and "Were the nobles and kings able to respect this, All things (*wanwu*) would be able to develop along their own lines."[7]

Thus *wuwei* manifests itself in the adaptive integration of the interacting parts or units of a complex system. The tendency of *wuwei* is to lead a system toward a state of *fractal congruence*, a close fitting together, a coalescing, of its significant components. The cooperation of those components facilitates the functioning of the system on its new holistic level. At the highest levels of complexity, *wuwei* integrates human beings into the systems of the world. This process of integration represents a synergistic enhancement of Dao by empathic human participants in cooperation with nature and with the outgrowths of human nature that have emerged from our evolutionary history and constitute culture. Thus, a fractal self is such a human being who is a participant with others, open to various worlds within nature and culture, and this person becomes a potential facilitator of emergence.

Furthermore, we suggest a fractal self is capable of growth and a kind of metamorphosis. Daoists refer to a seasoned human cooperator and facilitator, working adroitly, with a natural ease in the smooth, orderly, adaptive spirit of *wu-*

wei, as a sage. Such an individual is typically embedded in a particular *affinitive complex system.* An affinitive system is virtually anything in nature or human endeavor that is avidly sought by an individual in pursuit of vocation or avocation—a business, social, educational, artistic, scientific, or governmental enterprise, and so forth. Such systems typically develop chaotic structures and behaviors; envisioned as geometrical forms, they often constitute complicated attractors; around the edges of their coherent existence they would tend to be fractally organized, transcending classic dimensionality (see introduction). The sage tends to develop into a leader or catalyst within his or her affinitive system as he or she progressively "evolves" over time into increasing levels of intimacy and coherence with the system. We will have more to say about the astonishing potential of the fractal self as sage in chapter 8 and as a leader in chapter 10.

The Flow of Dao: Approaching the Edge of Chaos

In complexity theory, self-organizing systems generate new interactions or connections among their units that may be as variable as atoms forming protein molecules, species in ecosystems, or people forming institutions and societies.[8] As they develop, systems tend to evolve toward what has become known as the edge of chaos, as we noted in the introduction. Ample evidence has shown that the edge of chaos is a unique state of being in which a complex system functions at its highest level of dynamic activity while still maintaining its structure and integrity.[9] The edge of chaos, also known as a deterministic limit to orderly structure and behavior, is a precarious realm, however. On one side lies deep chaos, in which functional order disappears in rampant entropy; randomness destroys a system's integrity, and its units, and perhaps the system itself, may be extinguished. On the other side is a region that complexity theorists call the frozen realm, or the dwelling point of stasis. Here richness of connectivity diminishes, and a system retreats into immobility; order becomes rigid; dynamic flow or activity of the system's units ceases.

There are numerous textual references that suggest Daoist thinkers metaphorically visited these realms long ago, having become aware of fractal nature's shaping of the edge of chaos along what modern theorists refer to as the transition to turbulence. Perhaps nowhere is there a better model for the edge of chaos than at the interface of yin and yang that emerged from the formative philosophy of the *Yijing* we mentioned above.

Reflecting back to our discussion on *mana* in chapter 2, we now find the Chinese counterpart to *mana* in the form of *qi,* the psychophysical energy that flows through all things. Yin and yang are the two aspects of *qi* (氣). Chinese philosophers, in their articulation of the processes of the world, discuss everything as an aspect of everything else. In a profound sense, Dao doesn't flow through all things, but is the flowing itself, and all things in the world are interconnected with each other mutually influencing within Dao. Complexity theorists would envision a continuing flow proceeding through self-organization to myriad possibilities of emergent order. Through the interaction of yin and yang, order arises among the myriad things (*wanwu* 萬物), that is, "the manifold of everything that is happening,"[10] as they are blended in the world. The region where yin and yang meet is the precarious, dynamic edge of opposing tensions, where the flow of Dao inexorably leads, as when a water molecule cycles through the motion of wave formation up to its transition as breaking surf.

The region where yin and yang meet is drawn more fractally than the way the Taiji (太極) symbol (often referred to as the yin/yang symbol) typically portrays their meeting (figures 7.2a and 7.2b). Yang originally referred to sunshine and yin to shade. There is a detectable order to the process of shade overcoming sunshine—and sunshine overcoming shade—with respect to the valley between eastern and western mountain slopes, the natural loci of yin and yang.[11] In nature, this dividing line between yin and yang is a classic fractal border—the silhouette of tree branches and foliage, a canyon ridgeline, or a coastline. It is a dynamic line that changes constantly; this dividing line can never be simply circumscribed like the graphically portrayed curved line of the Taiji symbol.

Emergence (The Power of Dao)

Dao flows through the world in many distributaries and animates myriad systems, giving them the power (*de* 德) of *wuwei* (system structure). Their *de* is also referred to in the tradition as the *qi*, or *ch'i*, that flows through all things. In keeping with the model of self-organization in modern thinking about complexity, we suggest that *qi* is a way to refer to the organizing principle—the as-yet-unfathomed constructive mechanism behind universal evolution. In

Figure 7.2 In nature, the boundary of yin with yang is complex, fractal, and changes with time.

this context, *de*, the power of *wuwei*, might be characterized as the power of emergence. Both emergence and the essentiality of the holistic viewpoint seem to be clearly anticipated by Daoism. Zhuangzi, a later Daoist thinker from the end of the fourth century BCE, says, "Now the fact that when you point out from each other the hundred parts of a horse you do not find the horse, yet there a horse is, tethered in front of you, is because you stand the hundred parts on another level to call them 'horse.'"[12]

Seeking balance through tension, systems might be viewed as evolving through *fan* (反), the circular movement of Dao, also translated as "turning back." Two fundamental aspects of *fan* are the interplay of the opposites of yin and yang and the continuous re-turning, or oscillation, of the myriad things (*wanwu*) back to their source. When a particular phenomenon increases and reaches its terminus, it necessarily changes direction and recedes toward its other side. This progression from opposites is not coercive; it is natural and recursive, for it repeats itself in a patterned way yet retains its particularity, just as no two sunsets are the same and retain their uniqueness, yet they remain similar to each other in their familiarity. Each particular (myriad) thing realizes its self-nature (*ziran* 自然) through this process. The myriad things arise and are shaped by the interplay of the generative forces of yin and yang. Through shifting balances of these forces, the myriad things change their forms. In other words, through the blending of the forces of yin and yang along the edge of chaos, the self-generating emergence of reality, of nature, occurs.

As the reader recalls from the introduction, behavior of complex systems may be described (or circumscribed) by patterns known as attractors. An attractor may be viewed as representing the system's region of coherent existence.[13] Elements or units that are participants in a system interact within the often intricate geometry of the system's attractor. As we noted above, the realm that lies outside the attractor of a given system is the nether space of deep chaos—a region approaching randomness, in which the system loses its coherence and identity. The border of the attractor, however, is the system's most dynamic region. As noted, this region approaches the edge of chaos. Along this edge, turbulence begins to appear in the behavior of the system's units. Entering this transition to turbulence, elements of the system are able to "explore" the realm of near chaos, from which they may return to more predictable patterns safely within the attractor's borders. Flirting with the edge of chaos, however, can be viewed as risky for a system, owing to the chance of disintegration into randomness and perhaps extinction.

Numerous discoveries in nature and in computer simulations of complex systems suggest that the edge of chaos is also the prime locus for emergence.[14] Emergence often appears as "transcendence." The new pattern or entity is more than the sum of its formative parts or actions, such as Zhuangzi's "horse." Emergence often appears to inflate from rather small regions of turbulence along the edge of chaos.

Dao is variably characterized in the *Laozi* as an easy-flowing stream that can run in any direction (chapter 34); returning (chapter 40); crooked (chapter 45);

perfectly level, straight, and hence easy to follow if one knows the way (chapter 53); indefinite and vague, nebulous and dark (chapter 21); nameless and unworked (chapter 32); and empty and abysmally deep (chapter 4). These attributes may seem to be mutually contradictory, but could readily be ascribed to different parts of a complex system's attractor and its border regions.

The oscillations of Dao give rise to *wanwu* (the myriad things or everything that's happening) in the space between yin and yang. As conceived by authors of the *Laozi*,[15] the world takes shape in the turnings between the determinate (the having, the being); and its opposite, the indeterminate (not having, nonbeing), thus evoking images of *waves* of emergence—stirrings in quantum foam, the interchangeable nature of matter and energy, the creative turbulence between male and female—new worlds bubbling up at the edge of chaos. According to the circular movement of Dao, or *fan*, the myriad things are capable of changing their forms to generate new patterns in the world.

> Everything carries *yin* on its shoulders and *yang* in its arms
> And blends these vital energies (*qi*) together to make them harmonious
>   (*he*).[16]

The Chinese word for "blends" in the above quote is *chong*. *Chong* (沖) has the vitality of gushing forth, a process that bespeaks emergence. This gushing forth proceeds out of the fractal harmony (*he*, 和), or pregnant order, and is possible only when there is yin and yang intersecting to form one.[17] However, the harmonious order that emerges is a dynamic one; this new order is one that will transform continuously and mutate in immeasurable ways.

Metaphorically, Daoism anticipates something like the butterfly effect.

> Way-making (*dao*) gives rise to continuity,
> Continuity gives rise to difference,
> Difference to plurality,
> And plurality gives rise to the manifold of everything that is happening
>   (*wanwu*).[18]

In Chinese thought, this transformation is called Dahua (大化), the great transformation, a principle that seems to express glimpses of universal evolution. Prominent Daoist scholar Tu Wei-ming writes, in his "The Continuity of Being: Chinese Visions of Nature," that

> the organismic process as a spontaneously self-generating life process exhibits three basic motifs: continuity, wholeness, and dynamism. All modalities of being, from rock to heaven, are integral parts of a continuum.... To say that the cosmos is a continuum and that all its components are internally connected is also to say that it is an organismic unity, holistically integrated at each level of complexity.[19]

This great transformation is self-organizing and continually manifests throughout the continuum from earth to sky—from the rocks to heaven. Tu further discusses this spontaneously self-generating life process in terms of *qi:* "The continuous presence of *ch'i* [*qi*], the psychophysical stuff, is everywhere.... The continuous presence of *ch'i* in all modalities of being makes everything flow together as the unfolding of a single process. Nothing, not even an almighty creator, is external to this process."[20]

Daoism's vision of nature places humanity on the inside of the unfolding of the self-contained and containing system we call the world. As we shall see, this relationship is a hopeful one that suggests human beings have a potential for vastly extended participation in the universe. As the *Zhuangzi* states, "The Way comes about as we walk it,"[21] that is, with human interaction and participation. Daoists have always realized the inexplicable magnificence of the universe, the world, and the depth to which humans are capable of cooperating with the natural world. Dao anticipates an essential evolutionary-cum-revolutionary participation by human beings, aspiring to become fractal selves whose future may catalyze great transformations as a butterfly rises toward the heavens and triggers a changing weather pattern, or perhaps even more.

Interdependent Arising

While the Daoists developed an elaborate worldview that anticipated self-organization leading to complexity, the edge of chaos, and occasions of emergence, philosophers of a different Asian tradition cut through the world-shaping process to focus closely on the depths of emergence that come about through the interdependent tendencies of nature and human nature. We see in Huayen (Chinese) or Kegon (Japanese) Buddhism that they would especially make significant use of Indra's Net, which we suggested in the introduction as a representation of the structure and order of the universe. For these two schools, Indra's Net vividly represents a number of prominent Buddhist ideas, such as interdependent arising; the non-dual nature of all that is, *anatman;* and the interconnectedness and relatedness of all things. In the *Avatamsaka Sutra*, Indra's Net is given vivid expression as a vast web stretching out infinitely in all directions, with

> a single glittering jewel in each "eye" of the net, and since the net itself is infinite in all dimensions, the jewels are infinite in number. There hang the jewels, glittering like stars of the first magnitude, a wonderful sight to behold. If we now arbitrarily select one of these jewels for inspection and look closely at it, we will discover that in its polished surface there are reflected all the other jewels in the net, infinite in number. Not only that, but each of the jewels reflected in this one jewel is also reflecting all the other jewels, so that there is an infinite reflecting process occurring.[22]

One of the best ways to visualize Indra's Net is through the scientific advances brought about through holographic imaging.[23] The technique of holography enables the three-dimensional image of scattered light from objects to be reconstituted and reconstructed when the original source is no longer extant.

Given their sense of interdependent arising, which we will soon discuss, Huayen and Kegon Buddhist thinkers realized, perhaps more strongly than the makers of the original myth of Indra's Net, that anything done to one part of the whole affects everything else, since everything is ultimately interconnected with everything else. In other words, if we pluck a jewel from a portion of the net, a hole ensues, leaving the rest of the net affected and ultimately debilitated. "Holographic" means to somehow write the whole. In writing the whole, we need to speak in metaphors such as Indra's Net, or even in the language of science. In Buddhism, especially in its depiction of the universe as exhibited in the image of Indra's Net, this image becomes most relevant to our thesis, as giving expression to the way of the fractal self.

Buddhism's probing of reality, with its strongest expression in Zen, emphasizes the world of pure experience. Mere concepts of experience can never stand for the real thing; conceptualization of the world leads to a false, disconnected realm of static isolation, wherein the self is no longer a collateral and interdependent being. The power of conceptualization in its highest abstraction leaves an impoverished self without an embedded place in the world. The Zen adept practitioner (like the Daoist sage) enters into direct communion with reality and the objects of reality that compose the flowing, shifting, complex systems of the world.

At one level, Buddhism embraces a sensual way of being, and Buddhists seek to participate in the world directly and immediately through their senses. To take a simplified example, when we have a cup of tea, we have a direct experience of tea if we are mindful of sensations at the moment of drinking. This immanent experience of tea is wholly different from a conceptual experience of tea. Later, if we reflect on the experience of drinking the tea and isolate the experience as being distinct from our other life-experiences, the original direct sensory immersion (the appreciation of the warmth, aroma, and other qualities of the tea's blend and preparation) is rendered illusory in the sterile realm of concept that lies outside of existence. To a Zen adept, at the moment of tasting the tea, there is no distinction between the qualities of the tea and the drinker of the tea. Only when a distinction is made between subject and object does the experience of this small part of the world disappear.[24]

The development of the dualistic relation between self and world or subject and object (a product of the mind/body problem of Descartes) is at the root of issues as disparate as our aesthetic judgments about art and the environmental crisis. Thus Zen's posture of non-duality in typically more complex realms than tea drinking represents an anticipation of cooperative self-organization in nature (the merging of self and other in a complex system) and is also an aesthetic approach to the interbeing of all things in the world. The Buddhist doctrine known in Sanskrit as *pratītya samutpāda*, which is usually translated as dependent origination, is really a doctrine of interdependence, that is, the interbeing of all things.[25] *Pratītya samutpāda*, which is found at the core of all schools of Theravada and Mahayana Buddhism, denies the ontological status of any separate, isolated, or distinct being defined in independent essential or substantial terms. Rather, everything participates in intermeshing systems that shift in their qualities and manifestations of integrated structure and behavior—moment by moment like cloud patterns in the sky, or through eons like microbes in the primordial seas. Thus, Buddhist life is steeped in intimacy. All being is interbeing, and even Being, the most universal of all concepts that transcends any categorical distinctions made in apprehension of the world, is ultimately only interbeing. Even the human self is not really a self from this perspective. The self is really a not-self (*anātman*), that is, a being that from necessity emerges and is conditioned constantly by other beings in the field of all interbeing. From this perspective, the self does not exist in any ultimate sense as a separate entity.

Pratītya samutpāda can also be translated, and more appropriately, as "interdependent arising" and ultimately means that everything is impermanent and constantly changing.[26] This Buddhist doctrine of impermanence is called the *anitya* doctrine. In this doctrine, there is the sense of endless possibilities of emergence at any given moment. We suggest this sense points to the interaction and "inter-experience" of components of complex systems as ever creative of the potential for novelty in the world. The impermanence and constantly changing nature of everything in existence leads Buddhists to the idea of *śūnyatā*, or emptiness. Since everything and any given entity in the world is in this radical flux (a type of change very different from Aristotle's idea of causation discussed in chapter 2), there is no abiding reality to anything. Our idea of such an abiding reality is only a conceptual projection placed upon an ever-shifting flow of reality. The prominent Mādhyamaka philosopher Nāgārjuna is very clear on this matter when he proclaims that "emptiness is interdependent arising" (24.18 *Mulamadhymakakarika*). Hence, from the Buddhist perspective, the false affirmation of a separate reality of individual identity—whatever condenses out of our conceptualizing after immediate experience—is the greatest illusion. This illusory way of being in the world is a cause of *dukkha*, or suffering (perhaps similar to unfitness, or maladaptive-ness in a Gaian ecological sense). And the affirmation of the not-self (*anātman*), that is, a self that is an interbeing emerging from the dynamic play of the conditioning forces of the unfolding world, is to eliminate *dukkha* and avow life. A healthy, progressive world embraces the human immersion in complexity and cooperation that maximizes the chance of emergent moments. By definition, we're there (here) already, especially if we assume a more intimate way of visualizing self and universe.

San Francisco–based photographer Nathan Wirth visually captures this sense in his Zen photographs. In the first of his images (figure 7.3) we can see the intimate immersion of the self at the pinnacle where sky and earth touch. The middle image (figure 7.4) reveals the emptiness (*śūnyatā*) from which all things, *the wanwu*, emerge; in the last image (figure 7.5), the self is integrated into the current of earth's flow with the subtle similarity across this scale revealed in re-iteration throughout all three images. (Also see chapter 9 on Dōgen's idea that the Buddhist *sūtras* ("threads," scriptures or discourses) are the entire universe and the enlarged sense of self and truth.)

As the Vietnamese Zen master, poet, peace activist, and Nobel Peace Prize nominee Thich Nhat Hanh says of Zen masters, "Their words, acts, and looks also have the function of combating concepts, of producing crises, and of creating

Figure 7.3   Figure 7.4   Figure 7.5

conditions that arrive at releasing the vision of reality."[27] Zen masters have developed numerous strategies to assist practitioners to have a direct perception of reality, one unhindered by human conceptual overlay. The Zen experience of directly perceiving the world abolishes our overly developed dualistic tendencies. Unlike the process of conceptualizing, the act of perceiving in the participatory moment contains both the subject and object simultaneously. In this realm and in this moment emergence is potentiated in the world. As Hanh notes in another revealing example:

> When the eye is opposite a flower, one can say that the eye and the flower are dharmas that can exist separately; but when *"seeing" occurs, the subject and object of the seeing exist at the same time in sensation.* The flower is not merely the object seen. The object of seeing is found in the seeing itself, and cannot exist independently of the subject of the sensation. . . . [W]hen sensation occurs, the first phase, which is contact between the physiological phenomenon (eye) and the physical phenomenon (flower), has already passed in order to arrive at the second phase, which is the sensation (seeing).[28]

This is the moment of emergence without the discrimination of subject/object, the dharmic fusion at which reality is revealed.

In so many ways, we manufacture our worlds, at least as far as giving our worlds structure. Sometimes this is accomplished through our metaphysical beliefs, such as our preoccupations with God, Being, the mind and body, and free will. Once we speculate on God, or develop universal concepts such as Being, or the relation of mind and body, or on our volition in the world, we now have a metaphysical problem; we invent our philosophical problems, and in the past, and perhaps now, they drive our science and even our consideration of our perceptions. When we perceive an object, it is our retina (located at the base or back of the eye) that sends electrical impulses streaming to the visual cortex of our brains. Without the visual cortex part of the brain, we would not see. However, this part of the brain "sees" no light and remains for a complete lifetime in the dark, a darkness such as what Plato's cave prisoners experience in the beginning of Book 7 of his *Republic,* as they perceive the shadow reality in front of them only because of a light they never get to see. The retina itself, without the visual cortex, doesn't perceive anything either, since it is just a processor of the information sent to it from the photon field of vision in front of the eyes.

Some of the mystery of the sensual self's perceptive state of being is evoked by Francis Crick in *The Astonishing Hypothesis*.[29] Crick explains the use of various optical illusions and misperceptions as tools in neurological and psychological research on visual perception. Vision, contrary to what many people imagine, is not analogous to a television set playing inside the brain. Just as Hanh suggests, neuroscience reveals that the "object of seeing" is an illusion. In Crick's words, "Thus, what the brain has to build up is a many-leveled interpretation of the visual scene, usually in terms of objects and events and their meaning to us. As an object, like a face, is often made up of parts (such as eyes, nose, mouth, etc.) and those parts of subparts, so this symbolic interpretation is likely to occur at several levels." Two paragraphs later, he concludes, "It is difficult for many people to accept that what they see is a symbolic interpretation of the world—it all seems so like "the real thing." But in fact we have no direct knowledge of objects in the world. This is an illusion produced by the very efficiency of the system since, as we have seen, our interpretations can occasionally be wrong."

The mutual cooperation between the retina and visual cortex gives us seeing, but what is this seeing that is seen? Perhaps the nature of reality—and we know a lot already from physics—indicates that the so-called reality outside of ourselves is equally complex. By themselves, our manufactured images of self/world/universe are something far from the practical sense we typically attribute to them, for they give us nothing but a fabricated world of self/other; and it is precisely this world that has been called into question by Daoist writers and, especially, by Buddhist thinkers after they come into interface with indigenous Chinese philosophy. Combining the wisdom of Hanh and Crick, we could say that *feeling* as one with our brain's explicit symbolization of the world around us is as close as we can approach intimate human participation in the "real world." Both Hanh and Crick deny any little self, a homunculus, residing inside of us.

A Fork in the Stream

As we have seen, central reservoirs of thought within the major Asian philosophies of Daoism and Buddhism collected ideas that coalesced notably in holistic models of the world. The essentiality of human cooperation with systems in that world of given complexity comes through strongly in these traditions—perhaps a legacy from preliterate ages (see chapter 2). In the West, some of the earliest Greek thinkers flirted with holism, but then, with a new metaphysical orogeny, an opposite worldview arose, gained elevation with tectonic force, and from it streams of thought began to flow through a widening watershed of reductionism that ultimately became the foundation of nearly all modern science. Subject versus object, observer opposite observed, self and other came to the fore.

However, in the Western account of the world, with its roots in the classical Greek period, there was a dominant tendency toward a kind of rationalism. Philosophy and science were always seen as one with a special reverence for reasoning from first principles. In part, as we have seen, even the ancients began their models of the world and universe in this way. And then, various creation myths sought to explain how complexity came into the world from associations or cooperations among the most basic elements discernable or imaginable at the time. This flow of thought appears to have initially entrained the ancient Greeks, who contributed deep and powerful currents, at first eddying around natural complexity and humanity's relationship to it, but ultimately flowing away from intuitive consideration of holistic properties and seamless relationships in the world and toward reductionism as the mainstream approach to understanding the ways of nature and human nature.

As noted in chapter 2, the earliest Greek philosophers envisioned the nature of the world as fluid. Water, air, and a kind of boundless ether were supposed as representing the nature of everything. Traditionally, Western philosophy and science begins with a fragment from Thales as cited by Aristotle that stated the nature of all things is water and that this fluid cosmos is ensouled and full of gods, dissolved like elements in the sea. We can see the influence of an earlier primitivism in Thales' idea that everything cosmic is divine, but in his vision that all is inherent in water we begin to see an example of a reflective mode of being (emergence from some ultimate beginning state of togetherness—flow connected to flow in space and time) on the part of the self. Following Thales was Anaximander, to whom the universe was *to apeiron,* the indefinite or boundless. Anaximander appeared to have looked at all that is and saw a world similar to that of Thales: in it all is one interconnected whole. Anaximander's monism was less reliant on one particular element such as water and articulates a reality where there is ultimately a flowing similarity of all things across scale that defies any metaphysical definition. Herakleitos, following such fluid models of the world offered by his predecessors, imagined for the first time a self that could reflect on its selfhood, but that could be virtually anywhere in the world and possibly become part of that fluidity.

Herakleitos also saw a universe without definite boundaries, but he added a crucial dimension lacking in his predecessors' thinking. Herakleitos' universe was dynamic and constantly changing. In one of his most prescient fragments* he states, "One cannot step into the same river; nor can one grasp any mortal being a second time in succession, but swift and piercing it changes scattering and it gathers again, again and later, but at once it forms and dissolves, and approaches and lets go."[30] Herakleitos is known as the Weeping Philosopher, per-

---

*Only fragments or small portions of the texts of the Pre-Socratic philosophers remain, and hence the term "fragments." Often, these fragments come to us in the form of *testimonia,* or testimony, from other authors. Aristotle is one of the main sources for the early nature philosophers who preceded him. He typically will present their views in an attempt to refute them. Herakleitos frequently falls into this category.

haps because he realized that everything in the universe is part of the eternal flux and that there's nothing exempt from this impermanent perpetual process, as noted above with the Buddhist doctrine of *anitya*. There are no exceptions—even for the self, the soul, the human. For him, eternity is always immanent; in other words, it is always with the human (and everything else) in the flowing milieu of all that is. These momentary "gatherings" that we are, that all things are, are simply just that:

> Gatherings—
> Whole and not whole
> being brought together, being cast about
> In tune, out of tune
> From all things one
> And from one—all things[31]

This shifting universe, however, expresses its changes in orderly and emergent ways and retains the naturalistic divine characteristics inferred by Herakleitos' predecessor Thales. As Herakleitos recounts, "The god is day-night, winter-time-summer-time, war-peace, satiety-hunger [all the opposites, this is the sense] and 'he' is changed just as fire which when it is blended with that which is burnt [incense, spices] it is named according to the pleasures of each."[32] The tensions of these opposites create the opportunity for emergent moments to arise, but also introduce the ultimate fatal duality, if human beings take center stage and are taken out of nature—something Herakleitos could not envision. He does not fall victim to the later Greek dualism of the Pythagoreans or Plato as he again insists with his statement that "if you have listened not to me but to the *logos*, then it is wise to agree that all things are one."[33]

Herakleitos' divinity is also subject to the same natural processes as all things in nature are—god *is* the coming-to-be of day and its passing into night and as night comes-to-be day it passes, and so on across the scale of daily cycles, solar cycles (winter and summer), social and political cycles (war-peace), and our individual turnings of satiety and hunger. The emergence of all that is in our world is seen as the synergistic unfolding of the universe, and this too is considered divine, for his god is all of these progressive and mutually inclusive opposites across scale. These sentiments are echoed in another remaining fragment: "This cosmos is the same for everything, not any of the gods nor men has made, but it always was and is and will be; an ever-living fire, kindling in metres and quenching itself in metres."* Herakleitos was too astute not to realize that change

---

*The Greek term μέτρα, translated here as metre, is the Greek root of the English "meter" or "metre" and means "that by which anything is measured" and applies to both liquids and solids, with other meanings including proportion and limit. Here a reasonable interpretation of Fragment 30 is that the universe is natural and accessible to human understanding through quantifiable measures. H. G. Liddell and Robert Scott, *An Intermediate Greek-English Lexicon: Founded upon the Seventh Edition of Liddell and Scott's Greek-English Lexicon*, 7th ed. (New York: Oxford University Press, 1945).

is not always completely orderly and that random events, if situated in alignment with other aspects of the system, can cause interruptions in the expected, orderly flow. Being open to the evolving universe and the possibility of the unexpected is essential because "a thunderbolt steers all things" (Fragment 64). Perhaps anticipating chaos theory's understanding of attractors and the systems in which change develops, Herakleitos demonstrates an intuitive sense very much like that of the Daoist sage (see chapter 8).

Herakleitos pushes beyond Thales and Anaximander even further. Even the self is viewed as constantly changing. Although he is the first Western philosopher to articulate clearly the emergence of the self when he states "I was seeking myself"[34]—perhaps he was breathing in the presence of the movement toward a self in the intellectual air of his day—he concludes that an individuated self, or soul, cannot ultimately be discovered: "You would not find out the ends of the soul, even if you travel down its every pathway; so deep a *logos* it has."[35] The complexity he glimpsed, not only in the changing world around him, but in the nature of a nonessential, embedded self in the very processes that constituted the world, led him along a pathway similar to the Buddhists. The self, or soul, cannot be discovered because it does not ultimately exist outside of the very processes that constitute it; these processes are not distinct and separate from "everything in the world that is happening" (the Chinese version of *wanwu*, as we saw above). We can note, Herakleitos' direction is not only downward into the dynamic and immanent play of worldly forces, but also perhaps outward, evoking images of clouds and stars, "gathering and scattering, scattering and gathering." Herakleitos' reflection, suggestive of the infinitely iterative plunge into the famous Mandelbrot set, cannot take him to a substantial, immortal soul or even to a defined sense of self. Herakleitos' notion of the self should remind us of the *anatman* of Buddhism that Thich Nhat Hanh so skillfully articulates. Herakleitos' epitaph was the "The Riddler" or "The Obscure" for a reason—his thinking forces us into a new level of realization: that to find the self as being embedded in the world is the challenge.

To grow the soul back into all things is to meld with, be conditioned by, and create a deep sensitivity to the successive progression of all that is flowing and evolving around us. This growing soul—immanent in universal evolution— provides early Western glimpses of the fractal self.

Herakleitos stood on the East-West divide. Subsequent thinkers in the Western tradition would depart drastically from this self-within-world. Plato, whom Alfred North Whitehead would credit as being the author of Western civilization,[36] envisioned the self clearly defined and distinct from the world in which it dwells. The rest of the world becomes "other," and the self will achieve its merit by aspiring to perfection through a process of distancing itself from the world. With this defined self comes an anthropocentric view of a deity-directed universe giving the human perspective a privileged and prominent position. The self and its God become transcendent entities, above and beyond the world, and no longer comfortably at home in it.

Reductionism, which flooded the Western intellectual landscape after Herakleitos, is the philosophical and scientific way of knowing that identifies the components or units of systems of nature and seeks to understand their basic properties. This process of dissection proceeds backward and downward, approaching simplicity in ever more fundamental units and properties. Reductionism has been hugely successful in interpreting the structure and workings of nature from subatomic to cosmic scales. Nearly all of modern science and technology is rooted in reductionist discoveries. However, like classic geometry, reductionism is rigidly entrenched in discrete levels of organization. It cannot follow the butterfly that seamlessly amplifies its influence across scale and transforms systems it touches. Only recently has a clear view of this shortcoming of reductionism begun to emerge: arguably its greatest failing is that it offers no possible understanding of emergence itself.

Emergence, however, has its deepest roots in philosophy. After Herakleitos, it was restored to life in the West primarily by Dutch and German philosophers in the seventeenth and eighteenth centuries. Spinoza, Leibniz, and Kant each made contributions in their own ways. Spinoza is most famous for his *Deus sive Natura* (God or Nature), which triggered considerations of emergence in physical environments in Leibniz's work on fossils and his interest in Latin translations of the classics of Chinese philosophy, as well as in Kant's earlier thinking on the evolution of stars and planets and his biological reflections that organisms are "in the highest degree contingent" and "mechanically inexplicable."[37] Although these roots are indeed present in Western philosophy, much deeper crystalline insights into emergence can be found in Buddhist and Chinese thinkers.

From Self to Sage

> In all things the quintessential
>
> Transforming becomes the living.
>
> Below it generates the Five Grains [elements],
>
> Above becomes the constellated stars.
>
> Flowing between heaven and earth,
>
> Call it the ghostly and daimonic.
>
> Who stores it in the breast
>
> Call the sage.
>
> —The Guanzi

ormative Chinese philosophy would sprout into Confucianism, Daoism, and their later adoption of and adaptation by Buddhism. These emerging traditions envisioned a world that, redescribed in modern parlance, is composed of shifting, fractal dimensions in which emergence can happen along turbulent boundaries of attractors where tensions arise between opposites. For the ancients, including Presocratic philosophers in Western thought, these ways of thinking invoked nascent principles of our current understanding of deterministic chaos; they glimpsed self-organization proceeding toward complexity, and human beings moving through the world with a sense of participation. In this chapter and the next, we will see that emergent potential in a person leads to the highest expression of the self in Confucian, Daoist, and Buddhist environments where an effortless expertise may be achieved, arising from unforced participation within a particular sector or system of the world. The individual who attains such a state is recognized as a sage. Such a person is a cooperator in the broadest sense, very often an innovator and a catalyst and, in social systems, a constructive leader. Sagely behavior is perhaps the supreme achievement of biotic and cultural evolution.

As we have seen in the last chapter, Dao, the complex organizational pattern of the world, was built with an intuitive sense of interdimensional structure, uncannily resembling the forms that Mandelbrot uncovered many centuries later. And human beings are at their best when they are fractally embedded and engaged in this complex world, moving with and helping to shape its creative turbulence. We suggest that the sage is an ideal representation of the fractal self.

Furthermore, the authoritative (but never authoritarian) or consummate self of the sage seeks with inherent interest and attains by steadfast training a *fractal congruence* within an *affinitive system*. For the self-becoming-sage, fractal congruence means a natural fit or an approach to perfecting flair or fluency, and an affinitive system or attractor is the particular area of interest or niche in life to which one is drawn.

The Embedded Confucian Sage

Western continental philosophers, especially Nietzsche and Heidegger, believed that we live, in a profound sense, in our language. The language of the Chinese is very different from all Western languages and yields a radically distinctive sense of self as displayed in its pictographic and ideographic form of writing. It is considered by many that the legendary Laozi and Confucius were contemporaries, but whatever their chronology, they both continued the cultural tradition that had been passed down from their intellectual forebears of the late second and early first millennium BCE. The Confucian sagely self begins in the underlying linguistic model of what constitutes the highest attainable virtue. The human person, *ren* 人, was always seen as being produced by nature and forming a unity with the natural world. This view, according to Huang Chun-chieh, "supposes that 'person' and 'nature' share a substance and principle in common: *Ren* 仁."[1] The etymology of *ren* 仁, the highest virtue in Confucian philosophy, is usually taken to mean "relationality," which is the primary achievement of the sage. The word *ren* 人 (person) in this sense is a composite character of *ren* 人 and *er* 二, the number two.[2] Hence, this composite character (a self plus two) suggests that 仁 is attainable in relations of three or more, which is a perfect model for self-cultivation through the intimate and collaborative relationships between self and others because it echoes the most basic unit for the Chinese, the family: mother, father, and child. The family ultimately becomes, by extension, society writ large, which expands the field of focus or arena of action for the Confucian sage. As Roger Ames puts this sense of person,

> the timelessness and broad appeal of the teachings of Confucius begins from the insight that the life of almost every human being . . . is played out within the context of his or her own particular family, for better or worse. . . . In fact, in reading Confucius, there is no reference to a core human *being* as the site of who we *really* are and that remains once the particular layers of family and community relations are peeled away. That is, there is no "self," no "soul," no discrete "individual" behind our dynamic habits of conduct.[3]

In this Confucian concept, everything starts with the family. When a well-grounded family member is primed within a harmonious family structure and steeped in mutualism, he or she is poised to emerge on a larger stage with sagely attributes that may benefit community and society at large. Ames' notion here is that the family constructs the self, and this construction initially embeds the self in an interrelated world that can enlarge and amplify like a butterfly's wing vibrations that just may move across scale to higher and perhaps even more complex patterns and structures. The family is necessarily the cradle of the sage, and is an incubator for possible later emergence on a greater stage, and within a much larger family. Even the Chinese word that we might translate as "everyone" is *dajia* 大家, which literally means "big family."

Within this larger family, the self is faced with even greater challenges for self-cultivation. As we recall from chapter 6, Mencius (Mengzi, circa 371–289 BCE), a follower of Confucius' philosophy and revered by many in China as "the second sage," continues and augments the master's vision. In discussing Mencius, Huang Chun-chieh remarks, "People need to practice deep inner reflection in order to discern the content of *ren* [仁] and to reach the sphere that Mencius described as becoming 'sincere through self-reflection.' At this level, one realizes that the self and the other are not opposed or separate."[4] Our own interpretation— and of course this reflects back to the self's productive connection to family and community—is that this merging of self and other becomes a shared value for co-creativity and is none other than *ren* 仁, or the self's process of cultivation, allowing it to become a meaningful contributor to the evolution of society. According to Huang, this sincerity through self-reflection is an essential aspect for the *shengren* 聖人, or Confucian sage. In other words, this potential *ren* 仁 and the harmonious humanness in all of its positive qualities is the raw material we all possess that transforms selves into sages.

In his thinking on this process, Mencius develops his "doctrine of sprouts." In this extended biotic metaphor, he says,

> the heartmind [*xin* 心] in feeling [compassion] at suffering has the first [sprout] of consummate conduct [*ren* 仁]; the heartmind in feeling shame at crudeness has the first [sprout] of appropriate conduct; the heartmind in feeling a sense of modesty and deference has the first [sprout] of propriety [*li*] in conduct; the heartmind in feeling a sense of approval and disapproval has the first [sprout] of wise conduct. Persons having these four [sprouts] (*siduan*) is like their having four limbs.[5]

Mencius used the words *duān méng miáo* 端萌苗 for sprouts; literally, they mean a germinating plant. Thus he refers to our innate capabilities and tendencies as our starting-points for growth and the cultivation of our virtues in relationship with others. This directly reflects the conclusions of Frans de Waal about the evolved social cooperative tendencies in apes that proceed initially out of emotions. But like the growth of all sprouts, such developments in a person take time, applied energy, and "community support" (whether in nature or society) to manifest healthy growth into the future, much like a child's social and spiritual development through parental nourishing and familial self-cultivation. These sprouts of virtues are not mirrored from some distant beyond as they were for Plato or in the later developments of Christianity and Islam. They evolved in our primate ancestors and are the nascent dispositions in the self that are reflective of their social and natural world. Such inherent potential recalls how most anyone would respond to the fallen child in the well, as we noted in chapter 6.

In his prescription for developing and refining this natural empathic tendency, Mencius is pointing out the way a fractal self may begin to mature into a sagely being. Moreover, we are endowed with the capacity for the intimate and intuitive insights leading to morality. In our earlier thinking, now being enhanced in scientific terms, we are positioned to realize how our humanness is related to belonging to the world and universe in which we evolved, and that cooperation and collaboration need to be our ongoing rules of thumb. This is the realization of sage, as we have applied the term in Confucian thought. As Huang expresses it,

> the reason why the "great man" [sage, as we are using the term] is able to form such a unity is because his mind adopts *ren* [仁] as its essential substance. It is for this reason that he experiences no separation even from the plants, trees, birds, or beasts. Moreover, each person simply needed to rid themselves of their excessive selfish desires and to be less self-centered, in order to release and develop their inner goodness (based on *ren* or bright virtue); this would enable them to reach the sphere of forming a unity with the myriad things.[6]

Roger Ames' take on *ren* (仁) diverges somewhat, without any language such as "essential substance" that reifies and hypostatizes *ren*: "Yet *ren* does not come so easily. Far from being an essentially endowed potential, *ren* is what one is able to make of oneself given the interface between one's native, initial conditions and one's natural, social, and cultural environments." He also adds that "*ren* is first and foremost the process of 'growing' (*sheng* 生) these relationships to become a vital, robust, and healthy participant in the human community."[7]

Following Huang and Ames' insightful observations on the history of Chinese thought, we suggest that this expressed unity between self and others, including nonhuman species, lends itself to the moral sphere emerging from the empathic and resonates with Frans de Waal's thesis that empathy is an evolved capacity that is at the root of all morality. Moreover, this evolved moral sensibility *ought* to be at the root of politics, the creation of societies and cultures, social interactions, and the basis of our educational systems, for it is clearly the root of a Confucian sage.

The one who embeds herself and embodies herself is the sage. As Roger Ames explains,

> using the Confucian vocabulary, we might describe the evolving careers of members of the community from beginning as mere persons (*ren* 人) to becoming exemplary in their conduct (*junzi*) for their community through achieving consummate relational virtuosity (*ren* 仁) with other people. For only a few, by coordinating and embodying in themselves the values and the meaning that distinguish some epoch of human flourishing, they have the ultimate distinction of becoming sages (*shengren* 聖人), and as such, sources of enduring cosmic meaning. In Confucian philosophy, the expectation is that

human beings and the natural, social, and cultural worlds they inhabit must be full collaborators in a flourishing cosmos.[8]

This linkage yields the *shengren,* or highly evolved sagely being. In our view, the fractal self is the formative state in which a human being, with its sincere participatory approach to life, may emerge as a sage. Although few reach the highest level of sagely being, virtually anyone can be on this path of personal developmental outreach and cultivation and rise to adeptness within his or her affinitive system.

Wuwei Revisited

Returning to the Daoist tradition, as we began to interpret the self in chapter 7, the sage as one "who does nothing and nothing is left undone" is one of the most perplexing characteristics of the self's highest aspiration. What are we to make of statements such as the one who "wants to rule the world, and goes about trying to do so . . . simply will not succeed," or "The world is a sacred vessel, [and] is not something that can be ruled. Those who rule it ruin it; those who would control it lose it."[9] When people act in controlling and ruling fashions, they cannot act as sages because the sage must "do everything noncoercively (*wuwei*)" and when the sage "does things noncoercively" then "nothing goes undone."[10] The term *wuwei* (無為), especially in the Daoist context, emphasizes the orderly becoming of the world; it is the natural, unconscious, and nondirected action within a non-teleological process model of the world to which Laozi refers above. The sage effortlessly approaches, melds with, and becomes one with the flow of the world. The flow of the world (proceeding out of its seamless structure) is Dao in action, or "way-making." Dao then is understood as *wuwei:* Dao "really does things noncoercively, yet everything gets done."[11]

The emphasis for Daoist thinkers is always on the natural qualities or characteristics of Dao (道). Dao happens, but its happenings move in subtle ways that appear to be without purpose. Traditionally, this process of Dao must be preserved by the *shengren* (聖人), the sage, who is also referred to as the *zhenren* (genuine person) and the *shenren,* the spiritual or daimonic person.[12] Paradoxically, this would seem to represent an intrusion into the natural flow of the world. However, Daoist thought considers the sage to be fully immersed in the structure and flow of nature. This immersion accords with the principle of self-organization in the science of complexity, because the process has its own sense, its own intelligence, in which the sage contributes significantly but does not stand outside the system. A process that generates pattern (order from chaos) within a system, or part of the world, emerges from the interactions of the world's, or system's, integrated parts. This emergent raison d'être, or the tendency of nature to develop an integrating structure that directs its entire system, is Dao, or the way of the world: "The Tao [Dao] constitutes the regulating structure of nature."[13] This is in keeping with the formulations of complexity theory.

The image of the legendary Cook Ding in the *Zhuangzi* never having to sharpen his knife has become a popular metaphor for sagely action and systemic participation. Within the very humble system of a kitchen (perhaps chosen for its inconspicuousness), Cook Ding achieves perfect congruence with his art. He never experiences the need to sharpen his blade because he "knows" where the bones, cartilage, and other obstacles of a clean cut lie. The fluidity of Cook Ding's (non)action is a result of his knowledge, which is the culmination of years of disciplined practice and the realization that he is embedded within the structure and process of his world, and one with its myriad creatures/things, and with Dao itself. His actions are fractally matched to the shape and flow of his world. Thus, Cook Ding

> was carving up an ox for King Hui of Liang. Wherever his hand smacked it, wherever his shoulder leaned into, wherever his foot braced it, wherever his knee pressed it, the thwacking tones of flesh falling from bone would echo, the knife would whiz through with its resonant thwing, each stroke ringing out the perfect note, attuned to the "Dance of the Mulberry Grove" or the "Jhingshou Chorus" of the ancient sage-kings . . . For the joints have spaces within them, and the very edge of the blade has no thickness all. When what has no thickness enters into empty space, it is vast and open, with more than enough room for play of the blade. That is why my knife is still as sharp as if it had just come of the whetstone, even after nineteen years.[14]

And when Cook Ding encounters a snag in his free-flowing technique, he informs the King in typical sagely style with these words:

> Nonetheless, whenever I come to a clustered tangle, realizing it is difficult to *do* anything about it, I instead restrain myself as if terrified, until my seeing comes to a complete halt. My activity slows, and the blade moves ever so slightly. Then all at once, I find the ox already dismembered at my feet like clumps of soil scattered on the ground. I retract the blade and stand there gazing at my work arrayed all around me, dawdling over it with satisfaction. Then I wipe off the blade and put it away.[15]

There are endless examples of *wuwei* in our everyday world that are similar to Cook Ding's flawless art in the kitchen. Consider an expert surfer such as the late Rell Kapolioka'ehukai Sunn, who achieved perfect poise in a massive wall of moving water and fleetingly affirmed and enlivened its form and power, or a legendary sports figure like Michael Jordan whose mastery on the basketball court shaped the structure and focused the energy of the game. Each in her or his own right is similar to Cook Ding, because each, like him, performs his or her action so skillfully and appropriately. Such action appears to happen without effort because it is so fluid. The self, as sage, melds with the system.

Sages are cooperators and catalysts par excellence in human endeavor. In an affinitive social system they tend to bring out the best in all the various people engaged in a particular enterprise. In a sense a sage, himself or herself, can become a powerful meta-attractor. In a small country far from any hallowed center of the cinematic industry, Peter Jackson brought together a huge ensemble of experts as well as talented neophytes from within New Zealand and many other parts of the world to create the *Lord of the Rings* films. They were not only actors but artists of many media—computer wizards, sculptors, painters, clothiers, artisans of all kinds, diversified musicians and vocalists—people who became essentially *whānau,* or extended family, with Jackson at its cooperative heart, an extended family intently focused on an extremely complex creation for up to several years.[16]

Filmmaking at its best often proceeds out of a fluid realm of self-organization with the kind of sagely guidance woven into the process by directors like Jackson. However, the same sort of emergent organizational wizardry can sometimes be found in more constrained circumstances. For example, the geologist Steven Squyres largely coordinated the scientific, engineering, political, and social spheres leading to the spectacularly successful explorations of Mars by the NASA rovers *Spirit* and *Opportunity.* Orbiting around and working within a government agency known for its orthodoxy and even rigidity, and at a time of notable agency failures, Squyres was everywhere the human catalyst—attracting the best efforts out of his large and hugely diverse team of colleagues in the planning, manufacturing, testing, landing, deploying, and driving, as well as analyzing scientific data of this ultra-complex enterprise.[17]

As the Earth moves, so does the sage. Cook Ding cuts the meat according to the placement of muscle and cartilage; the sagely player, in the zone, with his team flawlessly weaving their efforts with his, hits the fade-away jump shot within the context of the game; and the surf sage, attracting massive energy in perfect balance, shoots beneath the wave's curl in the eternal moment before the wave's break. The sage moves effortlessly and accomplishes tasks with ease and assiduity because he or she has reached an understanding of a particular sector of the world and its processes. Not only does the sage intuitively know the world, such a person becomes a part of it. It reflects oneself.

> Within yourself, no fixed positions:
> Things as they take shape disclose themselves.
> Moving, be like water,
> Still, be like a mirror,
> Respond like an echo.[18]

The Sage as Butterfly (Cooperative World–Shaping)

This analysis suggests that the self is transformed through its connection to the forces of the world. Through affinity and training, the self becomes the sage whose metamorphosis ultimately creates a participatory match within the structure of the affinitive system. This idea in itself is not revelatory, for others have also come to this conclusion. However, we extend our thesis to include the sage's impact on the world—how the world is transformed—through the sage's trained and disciplined, yet free-flowing, interaction with his or her affinitive system.

Figure 8.1 Zhuangzi dreaming he is a butterfly or a butterfly dreaming it is Zhuangzi. *Credit:* From an ink painting by Ike No Taiga, 1747–1776.

The sage collaborates with the "uncarved block," a common image representing Dao, in such a way as to optimize its possibilities through facilitating its potential transformations and not imposing any preordained order on the world. Sages are catalytic in their collaborations, and it is only in this way that they "carve" Dao, the possibility of emergent worlds and their immanent patternings.[19]

In performing their activity, sages may appear to be doing nothing because they are so attuned to the natural patterns of their specific attractor(s). In reality, however, the sage as butterfly transforms the world by making the finer distinctions and discriminations appropriate to the flowing of Dao.

As the sage achieves participatory integration with the system, the carcass becomes dinner worthy of a king, the game becomes a signature event that may be long remembered, and even the wave may become legendary, having met for a moment the participation of the sage. This vision is wholly antithetical to the concept of mastery over a system, or of conquering nature, or of ruling over other people. What happens, rather, is an approach toward congruence or confluence; the sagely master achieves participation from within rather than an effect from without. The surfer in the curl of the wave, player in the zone of the game, the carver in the contours of meat or marble, all contribute a changed identity to the system. Something new appears in the world. The participation of the self fractally alters that self's system and shifts its flow.

Thus, sages do more than just preserve the world; in some important way they metamorphose within their systems; they effect certain changes in the fate of the meat, the flow of the game, or the breaking of the wave, and these changes make the world different in the future. There is a sense in some interpretations of Daoism that sagely practice is childlike and all we need to do is just return to a level of unreflective innocence. But sages do not merely recover the lost innocence of childhood, they learn the power of discrimination, of when and how to act. The discriminations they make are the finer discriminations that craftspersons make in the service of their ends: the master chef, the quintessential player, the skilled politician, and so forth. Sages make the finer discriminations as they approach perfect congruence with the system in which they participate. We concur with A. C. Graham, the distinguished interpreter of Chinese thought, when he says that the sage "is spontaneous from the very centre of his being."[20] Graham usually translates *ziran* (自然) as "spontaneous" or "spontaneity," but rather than an end, there is the suggestion of a new beginning wrought spontaneously by the sage's manifestation of ultimate craftsmanship. In this light, the spontaneous realization of one's dynamic essence (*ziran*) can lead to conclusions resonant with complexity theory's notion of emergence and universal evolution.

Our interpretation of sagely action involves close parallels to the butterfly effect. Recall that amplification of minute changes in the system caused by the butterfly is possible if the butterfly becomes congruent, or resonant, with the prevailing air currents at the scale of its wing-beats, and then if the system remains seamlessly connected from small scale to large. The sage's role in the system, like that of the butterfly and the sand grain (see introduction), is to contribute potential complexity that participates in and may alter the world. If perfect fractal congruence is achieved, amplification of the sage's participation is potentially infinite. Usually, however, something in the real world/system breaks the seamless dimensionality, and the butterfly effect is obscured in minor realms of chaos, but the sage as a fractal self must always be sensitive to the possible amplifying complexity of his actions by doing "things noncoercively (*wuwei*), . . . [treating] the small as great [and] the few as many, . . . [To] take account of the difficult while it is easy, And deal with the large while it is tiny." The sage realizes the "most difficult things in the world originate with the easy, [and] the largest issues originate with the tiny." Because "sages never *try* to do great things . . . [is why] they are indeed able to be great."[21]

In the context of the sage as butterfly, Zhuangzi, as A. C. Graham has remarked, is "interested not in the nature which a man inherits at birth but in the Power which he develops by intensive training."[22] In the text itself, Zhuangzi remarks,

> by the training of our nature we recover the Power. When Power is at its utmost, we accord with the Beginning. In according we attenuate, in attenuating we become Great, and blend together the twitters of the beaks. When the twitters of the beaks blend, we are blended with heaven and earth.[23]

This theme of blending expressed by Daoist sages is the "Inner Training" expressed in the *Guanzi* (see epigraph). Such discipline is to learn to subdue and overcome the ego so the self can become a being that naturally responds to the "other." This discipline is the discipline of learning to be creatively spontaneous and responsive. This kind of discipline is requisite in creating a self that learns to respond authentically to the ongoing spontaneous unfolding of the evolutionary prospect and to be capable of adroitly performing actions that can ripple throughout the system in beneficial ways. Such disposition and comportment are the potential power, the *ziran*, of the sage that manifests in noncoercive ways (*wuwei*) and ultimately provides the chance to participate in and contribute to the endless reshaping of the universe's potential transforming.

Back to the Edge: Between Intimacy and Integrity

Buddhists also traced a path to the attainment of sagely relationship with the world. To follow this path and glimpse the heights to which it leads, it is helpful to revisit our sense of participation, or fractal engagement within an affinitive system, by further considering the intimate self that we introduced in chapter 2 vis-à-vis the self of integrity. Recall that Thomas Kasulis uses these heuristic terms to discuss cultural difference, especially between Japanese and American cultures, but his insights are easily applied and adapted to any culture, or institution, such as a corporation or university, or even to subcultures.

As major roots of cultural difference, intimacy and integrity not only intersect to form an intercultural spectrum, but also resonate with complexity theory and the principle of self-organization. This realization lends new strength to holistic Asian perspectives on nature and human nature. Western science and philosophy have long exalted integrity and reductionism in their interpretations of the ways of the world. The emerging holistic perspective represents a sea change in thinking about the world and even incorporates inspiration from the mythmakers and shamans discussed in chapter 2. And from this insight something very crucial is born. In Kasulis' own words, "Integrity tends to think of the world as something external to be managed through knowledge" whereas "intimacy . . . tends to see the self and world as interlinking—the goal being to develop a sense of belonging *with* the world, feeling at home in it."[24] This type of thinking was not, however, entirely alien to the West before Kasulis' realization. For example, Novalis (1750–1822), the German poet, author, and philosopher, averred that even philosophy itself "is really homesickness, an urge to be at home everywhere." Later German thinkers of intimacy such as Nietzsche, Heidegger, and the French philosopher Merleau-Ponty also expressed similar views when they talked about "how the 'true' world became a fable" (Nietzsche), "the worldhood of the world" (Heidegger), and the "flesh of the world" (Merleau-Ponty).[25]

The glories of the integrity view are striking when we look at the accomplishments of Western science, medicine, and technology that have emerged out of the largely single-minded pursuit of knowledge and applications of knowledge. But we can also see its impoverishing and even threatening tendency, which risks diminishing and damaging human potential and our sustaining environment. The abuse of integrity becomes exemplified in the forms of righteousness, authoritarianism, and destructively competitive interactions of people having impact on individuals and human institutions, as well as the environment on local to global scales, aggrandizing "self" against the other. Manifestations of integrity, of course, are legion at the levels of individuals, but extend into the social realm across scale, where they emerge in more or less closed cultures—tribes and other ethnicities of different sizes and levels of organization up to nations, and also institutions of many sorts such as corporations, even including the institutions of modern science and engineering. The risk of an extreme integrity stance in the larger world then becomes a cultural xenophobia, or fixation within narrow ideological boundaries. Science, itself, has been accused of sequestering

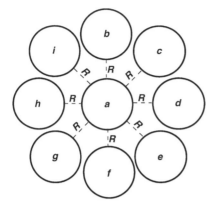

Figure 8.2

its culture in an "ivory tower." Groups of scientists may be drawn to fatal attractions such as research leading to progressively more powerful weaponry. In such cases, individual and cultural integrity may lead to hubris and the seduction of power-in-knowledge. Fortunately, as exemplified by the history of nuclear weapons development, not all scientists march in lockstep, and "integrity" among some prominent nuclear physicists in the mid-twentieth century also meant conscience in the minds of Robert Oppenheimer, Hans Bethe, Andrei Sakharov, and others who identified intimately with humanity on a global scale.[26] (We will return to these ideas more specifically in chapter 10.)

An integrity self defines itself as a free and completely autonomous being with a "free will" and with a profound choice and control in determining to whom and what it wishes to be connected.[27] The diagram in figure 8.2 might be helpful to illustrate the sense of the integrity self.

In the diagram, "R" represents the relationships of a my-self, "a." The relationships are external and correspond to the fixed boundaries of the ego. Although connected to others, the connections are external to the essential identity and are made by the assertion of free choice and personal volition.

What we are suggesting is that the fractal self can never be understood or realized from this integrity standpoint. Immersion into the free flowing systems of the world is necessary so we may embrace emergent order by engaging with the systemic flow of the interactions in the natural world, society, or universe. To accomplish this state, a different sense of self needs to come to the fore—the self of intimacy is needed, since it involves, as Kasulis states, a "necessary connection with others (either people or things)." The self of intimacy can be depicted as the diagram in figure 8.3.

The full circled "a" (my-self) in figure 8.3 is defined relationally, and rather than integrity's independence and external relation to everybody and everything, intimacy chooses interdependence and interconnectedness. This choice defines the self in relation to its surroundings.[28]

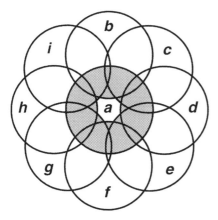

Figure 8.3

Throughout especially East Asia, we discover some of the strongest intimacy-based cultures in the world, and this tendency has led to sometimes incredible cohesiveness. But often this cohesiveness has devolved, not into harmony but rather into uniformity. For example, when it comes to matters of immigration and citizenship, Japan often displays a strong tribal sense of integrity. Intimacy is a very powerful force within Japanese culture, but when not extended on a more complex global scale, this force has had detrimental effects on Japan. Ironically, on a national level this tendency has become its opposite, a very integrity-minded standpoint with regard to other cultures and nations. This grip of excessive intimacy can be detrimental in the greater world, as when the first Japanese immigrated to Hawai'i. Their idea was to make money and return to the motherland, and the Japanese were rather clannish in their new home. A generation later, things changed some, but the Nisei, or second-generation children of the Issei (first generation), still resembled their parents in many ways. By the third generation (Sansei), however, the children displayed signs of strong integration as Americans. They lost their Japanese language ability; many social customs and traditional food preferences were transformed. Both intimacy and integrity models are emergent in their particular contexts and are not in any way inherent to individuals or groups of individuals. In chapter 10 we will see representative examples of extreme harm on the world developed more fully. The borderline and boundaries between intimacy and integrity is a complex one.

The difference between the two models of intimacy and integrity can be best seen in their extreme manifestations. According to Kasulis, the contrast between these cultural models is most vividly apparent in the differences between the Western self of existentialism and the Buddhist sense of self. Existentialism's emphasis on autonomous freedom requires the self to control external circumstances that reduces its standing as an individual. The goal for the existentialist is to not allow the facts of the external world to control the self; the greatest challenge becomes how to achieve a level of being in which autonomy can be

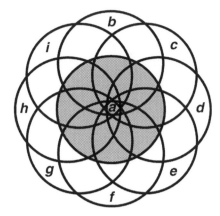

Figure 8.4

asserted over those circumstances and situations. Buddhism, on the other hand, "pushes the intimacy orientation to its furthest logical point."[29] A diagram of the Buddhist self shows the lack of any independent self in its most radical dimensions (figure 8.4).

From the diagram, we see how the Buddhist idea is even more strongly a self of intimacy, and may be the paradigmatic example of intimacy, since it is a being that is ultimately defined by everything else. This is the Buddhist doctrine of interdependent arising (see chapter 7), where the self is the sum total of its situation, "the nexus of a series of completely interdependent processes *(b-i),*" and "the processes of which I am part lead me to connect to further processes."[30] (The Buddhist idea of interdependent arising will be revisited in chapter 9.)

Self-reflection is indispensable to understanding what cultural model is being foregrounded or backgrounded in a particular given context. This topic of self-reflection and self-understanding is the focus of our next chapter, where we suggest that we take the challenge of Delphi most seriously: "Know yourself." But we do so with a twist and take a Buddhist turn: that is, to know oneself is to know oneself as "all-self," and, counterintuitively, this sense of self is where the Buddhist no-self doctrine leads. As thirteenth-century Japanese Zen master Dōgen advises,

An ancient buddha said:

The entire universe is the true human body.
The entire universe is the gate of liberation.
The entire universe is the eye of Vairochana.
The entire universe is the dharma body of the self.[31]

From Self to No-Self to All-Self

The world is not poor enough to oblige us to search for a God outside of it.

—Hölderlin, *Werke und Briefen*

People outside the way regard what is not self as the self. But what buddhas call the self is the entire universe. Therefore, there is never an entire universe that is not the self, with or without our knowing it. On this matter defer to the words of the ancient buddhas.

—Dōgen

The philosophical issue of self-knowledge, or problem of *knowing* one's relationship with the world, which we've been exploring in the last chapters in the Chinese philosophies of Confucianism and Daoism, is complemented by Buddhists thinkers. Buddhist sages consciously struggle against the problems arising from a self strongly devoted to the integrity way of knowing and being. The Buddhist sense of self, as we saw in the last chapter, becomes the paradigmatic model for what constitutes a self of intimacy. Out of intimacy, as we have discussed regarding Thomas Kasulis' work, one understands knowledge as residing at the interface of self and world.[1]

Self in a Fabricated World

Once this important insight about the true location of knowledge is realized, a Buddhist would likely conclude that there are no sets of universal philosophical propositions, scientific principles, or mystical beliefs that transcend human immersion in those approaches to knowing and, ultimately, to understanding truth. This means that our fabricated constructs for knowing ourselves and living in the world—such as the mind/body problem in philosophy and psychology, the pursuit of allopathic versus homeopathic or naturopathic medicine, and various accounts of fundamental scientific theories—cannot stand as purely objective, or scholarly, approaches to truth. From intimacy's perspective, such constructs could be seen as missing something about the interface of reality and self. Even our "universal" scientific understanding of evolution is beginning to blur, as a wave of technological creation with the potential to alter the biosphere and our very selves has begun to arise—from molecular medicine, agricultural genetics, and artificial human organs to cloned animals and approaches to the threshold of artificial microbial life.[2]

In Buddhist thinking, all is fine with the fabricated world as long as one culture, or cultural orientation, does not dominate others or as long as our ideas do not interfere in our direct experience of what constitutes the world. In the best of worlds (unfortunately far from what we see in our contemporary context), that thinking would constitute a sort of un-authoritarian, global state of mind. In so many ways, however, Western cultural domination has swept across the planet, and this has been destructive because of the strong integrity emphasis of the West and its stance of superiority toward the perceived "inferior" other(s). Then, too, our behavior toward nature also reflects our grounding on the integrity–intimacy spectrum. Until very recently, the prevailing Western outlook has been to view ourselves as supreme beings on the earth, to assert dominion and practice conquest of nature.[3] It is only gradually that many of us have come to the realization that a healthy biosphere is essential to our own survival.

Buddhist insights on these questions of our pathways to knowledge, and also how we transform what we think we know into prescriptions for action, may even have a parallel with the uncertainty principle in particle physics. When we see ourselves as apart from the natural or cultural systems we seek to understand, we encounter an inevitable uncertainty—a block to deeper understanding of those systems. The Buddhist "uncertainty principle" extends to the human study of culture and nature at all scales, not just that of subatomic particles. From the standpoint of strong integrity, we cannot help but interfere with and change the systems we try to study, because when we act as observers we are automatically outsiders, probing the system intrusively, disturbing and distorting its true nature and behavior.

Objectivity has been essentially the domain of Western science, as it developed on the high ground of integrity, but a form of objectivity also emerges, or evolves, in sagely practice in more intimacy-based cultures. In such cases, human decision-making can be objective and yet still remain personal—and this is precisely the epistemological value of availing ourselves of the intimacy way of knowing. Kasulis is helpful here as well, when he points out that "personal objectivity" can be seen in any sport, such as springboard-diving, gymnastics, figure-skating, and so forth, where judges use their intimate knowledge of the sport for arriving at evaluations of performances. There is typically a high level of consistency displayed in the assessment of performance by experienced (and unbiased) judges. As Kasulis suggests, such "judges' objective perceptions of subtle differences in the quality of style is not a knowledge that can be tested by just anyone."[4] This perception is objective because it has been developed over years of practice, performance, and participation from within the affinitive activity.

This more personal and affective-objective way of knowing is what is drawn upon by Buddhist sages, priests, and masters. Out of their discursive ways of thinking and knowing, Zen masters employ a number of techniques to move their pupils to more profound and immediate levels of understanding the nature of self and the world. One such technique is the koan. Many have heard of Hakuin Zenji's (1686–1769) famous koan, "What is the sound of one hand clapping?" From the objective world of integrity, such a question cannot have a logical answer in its most obvious context. But from the perspective of intimacy, the authenticity of the response can be gauged by the adept master. An intimately conditioned, yet personally objective, response is similar to what we saw in the context of the Daoist sage. Relying on his years of practice, Cook Ding cuts through the meat, cartilage, and so forth, but he also cuts through the rigid layers of representational and conceptual ways of knowing. This emerges from a deeper level of knowing, for his fractal adeptness at "cutting through" connects his being to the world's context at hand.

No Self Apart

To develop this sense of authentic intimacy is to interpenetrate a world beyond the dualism of self and other or the framework of an integrity self, wherein all relation is experienced as external. To engage in such interpenetration is to realize one is part of Indra's Net and its vast network of jewels that stretches out infinitely in all directions. Each jewel reflects the light of all the other jewels in the net, making no individual jewel a light unto itself. All the entities in the net are then seen as being interdependent. The English poet William Blake writes of this fractal congruence in his "Auguries of Innocence":

> To see a world in a grain of sand,
> And a heaven in a wild flower,
> Hold infinity in the palm of your hand,
> And eternity in an hour.[5]

Indra's Net illustrates that the universe is interwoven and that the human self is just one jewel among infinite others, simultaneously reflecting each other in the unbounded process of what is continuously happening. As an embedded being in the net of universal relations, the fractal self propagates a multidirectional reflection that may amplify across scale as it pulses through that individual's particular world. Many things arise in this network of relations, and the fractal self may well affect any number of cores of developing complexities as they begin to converge in a particular system. These complexities just may evolve if the right connections are in place. The idea is to engage the possibilities from within, even though one may be uncertain whether the effects will manifest and be propagated into the future. We might wish to see this as another manifestation of the Buddhist "uncertainty principle." A fractal self could be anywhere, at any level, in the expanse of Indra's Net and may play a role in the system's progressing.[6] Not only does such an interconnected self radiate new light into the world, it can potentially shape new cultural, social, and political realities as well.

As we saw in chapters 2, 7, and 8, this deeper connection of the intimacy self to the world at hand also applies to the Confucian side of the Chinese worldview, where we find sensitivity to immediate context in human affairs. An individual's position in society is conditioned by much more than competitive ability and applications of appropriate knowledge. In the *Analects*, Confucius says "Authoritative persons [consummate persons, or persons of *ren*] establish others in seeking to establish themselves and promote others in seeking to get there themselves. Correlating one's conduct with those near at hand can be said to be the method of becoming a consummate person."[7] Such correlation is fundamental not only to the Confucian social scheme and its "intimate community of praxis and knowledge,"[8] but also to the Daoists' focus on the self's becoming more naturally connected; it also is central to the Buddhists' approach to understanding the truth of the connectivity of all things within the conditionally changing universe. Such connectivity lies at the core of what it means to be an intimate being and a fractal self.

In comparing all major philosophical traditions, the Buddhist idea of the self is the most radical in its sense of intimacy. The strange-sounding doctrine of no-self, or *anātman*, is at the center of Buddhist thinking and practice. When the primal Buddhist doctrines of impermanence (*anitya*) and interdependent arising (*pratītya samutpāda*) discussed in chapter 7 are applied to the self, *anātman* is the result. As we have also noted, the Buddha taught that the opposing idea of a permanent, abiding soul is the main source of delusion and the cause of suffering (*dukkha*). The problem of suffering, its causes and conditions, are of the utmost concern for Buddhist practice and represents the core of the Buddha's teaching in the Four Noble Truths. Stated briefly, these four truths are (1) Suffering (*dukkha*) exists; (2) There is a cause for suffering; (3) This suffering can be eliminated or "blown out" (*nirvāṇa*); and (4) There is a pathway to accomplishing *nirvāṇa* and this is the Eightfold Path, namely developing an intimate frame of mind in dealing with view, intention, speech, action, livelihood, effort, mindfulness, and meditation.[9] In the Buddhist final analysis, this path alleviates the problem of suffering, chiefly the psychological need and intellectual demand for a soul. The Dalai Lama explains the Buddhist position on soul:

> If one understands the term "soul" as a continuum of individuality from moment to moment, from lifetime to lifetime, then one can say that Buddhism also accepts a concept of soul; there is a kind of continuum of consciousness. From that point of view, the debate on whether or not there is a soul becomes strictly semantic. However, in the Buddhist doctrine of selflessness, or "no soul" theory, the understanding is that there is no eternal, unchanging, abiding, permanent self called "soul." That is what is being denied in Buddhism.[10]

As we seek a permanent and distinct immortal soul contained within our bodies, or even an individualistic spiritual sector of mind, we suffer the greatest illusion.

Our illusory quest for a substantial soul, the self of integrity, then, is at the very core of the problem of suffering for the Buddhist. In overcoming this seemingly obsessive human need and demand for a self so unrelated to anything else in the universe, the Buddhist analysis takes a much different direction in its investigation of what constitutes a self—one that the Scottish philosopher David Hume would also take much later, in the 1700s[11]—and this examination will ultimately give us the self of intimacy.

When we investigate the self beyond the ego and personality, according to Buddhists, we find nothing but the five *skandhas* (literally meaning "heaps" or "bundles," and is usually translated as aggregates). In the Heart Sutra, it is stated that "The Bodhisattva Avalokita, while moving in the deep course of Perfect Understanding, shed light on the five *skandhas* and found them equally empty."[12] These *skandhas* are the material body and sense organs, sensations, perceptions, mental formations, and consciousness, which are loosely bundled together. There is no self beyond these aggregates, and their bundling is what constitutes our mistaken sense of self and deluded sense of being in the world.

In his translation of and commentary on the Heart Sutra, Thich Nhat Hanh provides a helpful analogy to explain the emptiness of self: the five *skandhas* are likened to rivers. A river represents each of the moieties: of form (body), feelings, perceptions, mental formulations, and consciousness. Each of these rivers is flowing within us, and each is "empty of a separate self." The meaning here is that "none of these five rivers can exist by itself alone. Each of the five rivers has to be made by the other four." In life, "They have to co-exist; they have to inter-be with all the others," like the various organs in our bodies and the air in the lungs that enriches our blood and so on.[13] These flowing rivers, the five *skandhas*, flow together within and through each self.

Imagine the monk or nun in a monastery, spending the day not only meditating and studying sutras, but sweeping and raking leaves, cooking and cleaning dishes, growing and harvesting crops. Initially, they may mindfully perform their functions, making coordinated use of the *skandhas*; however, the activities later become meditative acts in themselves, wherein the self then melds with its activity, where there is no longer a self that directs the performance of the chore, for the being is totally immersed in the doing. This original mindfulness is transformed as a doing without the mind's involvement. Achieving this state, one becomes *wuwei* (in the same context as a Daoist would view the effortless melding with an affinitive system, or life-attractor), and to Buddhists such activity is *nirvana*, a blowing out of the self. And it is this blowing out that marks the end of our suffering.

Poetry of the Intimate Self

A glimpse at the poetry of the famous Zen poet, Bashō, will help illuminate the no-self doctrine in Buddhism. As we can see in the sparse style of the haiku, chosen for its simplicity and directness toward the natural, the self is identified with the world in which it resides.

> The sound of hail–
> I am the same as before
> Like that aging oak.[14]

The back story of this haiku is that a homeless Bashō enters his new hut, which was built by friends and disciples after a fire had destroyed his old one, and the sound of hail striking the roof brings about the realization that he has always been and somehow remains part of all that is. Just like the aging oak, he too is aging; his mother has recently died and his father is already dead, but in this aging he is somehow "the same as before."[15] Although this is one of his earlier haikus, Bashō has already glimpsed that his self is no different from all the other myriad selves in the world of nature that stretch geographically and temporally throughout all that is constantly becoming. In expressing this, Bashō shows us a self of profound intimacy that is interwoven in the panorama of life.

Often, the idea of a distinct and separate self is transcended and replaced by wholly natural images, as we see in some further representative poems of Bashō.

> orchid fragrance
> from the butterfly's wing
> perfuming the clothes[16]

A literal translation of the poem would read "orchid's fragrance / butterfly's wing from / incense to do" (*ran no ka ya / chō no tsubasa ni / takimono su*).[17] In the last line, *takimono* means the fragrant wood and other things used for incense and can even refer to firewood. Bashō in essence discovered the butterfly effect. Perhaps, like chaos theorists, he chose the butterfly because of its subtle fanning of wings compared to other creatures? The orchid's scent, incense, or the fragrance of firewood transform a human's clothing into something novel. Shamans and sages seek just this sort of transformation in a natural way (recall our discussion on *ziran* in chapter 7): they alter the world through their direct participation and intimate knowledge. This brings the possibility of emergence. The clothes now have a fragrance, and there's no telling where that fragrance may lead—to a lover and generations of offspring, or to a whole new innovative series of events.

Although the word "clothes" in Bashō's poem is not in the original language of the poem, the translation suggests the Japanese practice of fanning incense over washed clothes to give them a fragrance. By portraying the self in terms of intimate connectivity among the orchid, butterfly, and the human, Bashō reveals how Buddhist practice releases the individuated self into a potential endless web of communion throughout nature and human nature.

But this is only half of the story for Bashō, for he urges us to see that the self is also channeled back into the natural process from which it has emerged because of its intimate relationship with the processes of the natural world. To have the butterfly transfer the perfume to the clothes requires human cooperation from an attuned self that can remain motionless. This recalls the Daoist sage of chapter 7, one "who does nothing and nothing is left undone" and Cook Ding's coming to a halt when he encounters "a clustered tangle" in chapter 8 when he realizes that it is "difficult to *do* anything about it." To be patient, to await the butterfly alighting and walking around the clothes and flapping its wings brings the modification of the clothes from a simpler state to a more complex one. The human and the butterfly must merge in the moment, become one together, and only then does the perfume have the potential effect to create something emergent in a human life that was not present without this intimacy with nature.

Coming Home to the Universe

We began, like everything that is complex, as stuff created in stars, and we have come to the point of achieving the knowledge and power to willfully create the stuff of stars and more. As a succinct description of the 1952 test explosion of the first megaton-range thermonuclear bomb, dubbed "Ivy-Mike," by US scientists at Eniwetok Atoll has it, "Momentarily, the huge Mike fireball created every element that the universe had ever assembled and bred artificial elements as well."[18] The yin and yang of such nascent capabilities in the human species—the destructive looming beside the constructive in our nature—requires of us both hope and vigilance. Leaving the butterfly on the wing, we might also focus hopefully with Bashō on water, a medium vital to all life. Humanity's emergence on a threshold of godlike powers places us as a frog poised to make a splash in an old pond of the poet's most famous haiku—

> The old pond
> A frog jumps in
> The sound of water.

In this poem, we see what we've been seeing for some time now: how widely or universally connected the enlightened self may aspire to be. Our claim should be clear now—this enlightened self is the fractal self, for the old pond we jump into is the universe. It's an old universe, at least from our perspective of time, for we are the freshly arrived. We've come with a special and accelerating capacity to creatively participate in the future of this great pond, a small *Homo sapiens,* but one, as it turns out, with immense potential to amplify the pond's rippling once we've consciously chosen to enter it and explore its very depths. We know so very little about this universe we call our home, and the questions for us now become, how do we become comfortable in this home, and what is the future of our species beyond the threshold on which we now stand?

With the Dalai Lama, we believe a revolution is called for, certainly:

> But not a physical, an economic, or even a technical revolution. We have had enough experience of these during the past century to know that a purely external approach will not suffice. What I propose is a spiritual revolution. My call for a spiritual revolution is thus not a call for a religious revolution. Nor

is it a reference to a way of life that is somehow otherworldly, still less to something magical or mysterious. Rather, it is a call for a radical reorientation away from our habitual preoccupation with self. It is a call to turn toward the wider community of beings with whom we are connected, and for conduct which recognizes others' interests alongside our own.[19]

We suggest this revolution aligns perfectly with the cooperative ethos in universal evolution. We've collectively arrived at a stage where we can jump into the universe, creating a splash with emergent potential. These ripples bring us perhaps full circle back to the potency of the earliest ripples of the universe at recombination, the first stage of universal "enlightenment" we noted in chapter 1.

However, the universe can still appear strange and indifferent to us, as Stephen Crane, the American novelist and poet, knew when he wrote:

> A man said to the universe:
> "Sir I exist!"
> "However," replied the universe,
> "The fact has not created in me
> A sense of obligation."

Although we have this huge potential, Crane reminds us we're not home free. We should pay attention to Crane's caveat and rein in our hubris and look carefully at the world around us. Ronald Pine echoes Crane's reminder in his *Science and the Human Prospect* when he says that "we should not get too uppity about this; the universe will do just fine without us."[20] Instead of thinking about ourselves in such grandiose ways and attuning ourselves to otherworldly hopes, we need to attend to the future very close to home, in its anticipation of ever-increasing occurrences of droughts, fewer resources, continued economic and social disparity, an intensifying deep fear of the other, and an abysmal alienation of ourselves. If we can somehow bring ourselves to this juncture, this moment of crucial choice and consideration of what is present at hand, then we may begin to see, perhaps for the first time in an authentic and real sense, that we as a species now stand at a new threshold, a new dawning of time.

As a species we have ascended to a seminal status. We are poised at the very edge of Bashō's pond. But if we reject those natural systems, processes, and cooperative instincts that brought us here in the first place, we'll turn our backs on the evolutionary progression that transported us forward as beings embedded within the world. We can thumb our noses at the cooperative constant and retreat still further into the unproductive and uncreative stasis of ultimate integrity—we can become anticonservation and destructively conservative in protecting our narrow self-interest at the expense of the home in which we live and the process that breeds complexity and fosters emergence. We clearly have this dark option, and many seem to be pursuing it, as we will see in the next chapter.

These realizations lead to a new interpretation of the classic concept of free will. This free will that now confronts us so urgently is not the age-old free will of metaphysical libertarianism, meaning the appearance of freedom given to humans by an omnipotent God who supposedly wanted individuals to be agents of choice. Our sense of free will comes closer to what Thich Nhat Hanh has stated in a number of contexts: "Freedom is not given to us by anyone; we have to cultivate it ourselves. It is a daily practice. . . . No one can prevent you from being aware of each step you take or each breath in and breath out." Awareness, and becoming aware, is the freedom of the fractal self; it is a freedom of intimacy and the realization of our potential to willfully remain within the evolutionary process and become the process' compassionate benefactor. To choose the opposite is to follow the ways of strong integrity and accede in the authoritarian drive toward domination over other species and members of our own. It's to choose the path of aggressive competition where we continue to unsustainably consume the resources of the earth, exploit each other, and adhere to authoritarian and fundamentalist religious practice. These paths are likely ways to self-destruction and extinction.

No other species has ever had this choice. Either way, we are destined to make a significant earthly difference; perhaps this difference is even cosmic, we cannot know. Hopefully, we are capable of applying that free will by following the model of the fractal self we've been outlining in this book. This self engages in cooperative enterprises in accord with and support of the evolutionary process and moves us to embrace the emergent potential that can arise from our ripples in the old pond.

A fractal self always finds itself in between its self and others, and here we mean all others, including animals, plants, and even those supportive landscapes and seas that make up the earth's biosphere. Finding oneself among all things is the starting point, the moment of blending, and the place of becoming embedded in the unfolding flow.

Yet the fractal self will also need some integrity to jump into Bashō's pool of intimacy. As readers recall, intimacy and integrity should not be regarded as absolutes that exclude each other; they are more nuanced in their complementary nature, and a balance needs to be struck, like the balance of yin and yang that gives rise to Dao. Robert Aiken suggests, in his commentary[21] on the old pond and Bashō's frog poem, that we find ourselves in twilight, a time that hints that human action is in the middle of the two poles of integrity and intimacy. The time of twilight; it is a magic time where a peaceful and quieting calm descends on the earth.

Aiken points out, in his *A Zen Wave: Bashō's Haiku and Zen*, that *mizu no oto* (water's sound) in the last line of Bashō's poem is onomatopoetic (just like the English "plop") and suggests the parody of the poem by Gibon Sengai (1750–1837) should be understood as being most instructive:

The old pond!
Bashō jumps in,
The sound of water![22]

Figure 9.1 The long-held Buddhist belief that the starting place for enlightenment is to embed oneself in the perpetual flow of all that is, vividly expressed in this photograph at Wat Mahathat, Ayutthaya, by Julian Bound. Several stories have arisen in the lore of this root-embedded Buddha, notwithstanding how it commemorates the relation of the Buddha rooted in the flow of nature.

Bashō has become the frog, is the frog, because all of us are potentially this frog, newly arrived at the edge of the ancient pond, a pond mossy at the edges "and probably a little broken down."[23] The sound of our splashing into the pond is an echo of all that resides between us and our larger world. We can cause profound effects through our human actions as they ripple outward into the unknown future. As a species, we have a calling from the universe that no other species has ever heard before.

In comparing Bashō's pond to the Bodhi tree under which the Buddha gained his enlightenment, Aitken in the long tradition of Zen warns against the trap of quietism—some kind of mystical union with all that is and a place of stasis, a frozen zone—when he writes that "remaining indefinitely under the Bodhi tree will not do; to muse without emerging is to be unfulfilled."[24] The enlightened self must enter the world, and this entering the world as an intimate participant is definitive of the fractal self. The choice is now ours.

Ultimate Compassion and Freedom

Contrary to some popular conceptions, Buddhism is not merely some kind of romantic and mystical union with the world. The Buddha was unmistakable and unwavering in his realization that the appropriate way of living needed to reach out to others as the self extends outward into the world. As he rose from under the Bodhi tree a week after noticing the morning star, the Buddha proclaimed, "Now when I view all beings everywhere . . . I see that each of them possesses the wisdom and virtue of the Buddha."[25] Our reaching out must include compassion for other selves, including nonhuman selves. We can see how this type of reaching out can take presence in the Buddhist approach to living fractally with a global sense if we listen briefly to Thich Nhat Hanh, the passionate, clear voice of living as an interbeing: in his *The Heart of Understanding: Commentaries on the Prajñāpāramitā Heart Sutra* we see how understanding our fractal connectedness is a prerequisite for living a life of compassion and understanding.

> If we look into this sheet of paper even more deeply, we can see the sunshine in it. If the sunshine is not there, the forest cannot grow. In fact, nothing can grow. Even we cannot grow without sunshine. And so, we know that the sunshine is also in this sheet of paper. The paper and sunshine inter-are. And if we continue to look, we can see the logger who cut the tree and brought it to the mill to be transformed into paper. And we see the wheat. We know that the logger cannot exist without his daily bread, and therefore the wheat that became his bread is also in this sheet of paper. And the logger's father and mother are in it too. When we look in this way, we see that without all of these things, this sheet of paper cannot exist.[26]

Thich Nhat Hanh knows there is so much more in this sheet of paper than what he says in a short paragraph, for there is the cloud and its rain in the sheet as well as the parents of the parents of the logger and there is the bubbling of that ultra-dense plasma gas that started the universe and built the nuclei that captured many of the photon-retarding, free electrons that formed the first true atoms. Those nuclei, their electrons, the fleeting photons, and all the emergent properties of our extant universe are present in the paper. Looking more deeply into this sheet of paper "we can see we are in it too. . . . So we can say that everything is in here with this sheet of paper. You cannot point to one thing that is not here—time, space, the earth, the rain, the minerals in the soil, the sunshine, the cloud, the river, the heat. Everything co-exists with this sheet of paper. . . . 'To be' is to inter-be."[27] Hence the question for us in the twenty-first century is not the existential one of the past, "To be, or not to be." Our question is a different one, and even more challenging, "To inter-be, or not to be." Our choice is no longer just an individual one pertaining to our self-interest for a meaningful life, but a special one not only of and for our species, but for all species and the process of universal evolution. It is a worthy task, a great burden for our species perhaps, but certainly our greatest challenge as a species to seek a future through the disciplined application of our free will.

As we've seen, this freedom defines the choice between the will of the ego and the will of and for the whole. Choosing to remain in the cohesive evolutionary process is to define the self as an internal participant among many others. A type of sagely partnership arises when the fractal self is embedded in the flow of nature and compassionately concerns itself with being a part of that momentum. This process is a divine one, but not with God as Creator. The sacredness of this process is the miracle of life itself through the affirmation of the fortuitous occurrences that present themselves at junctures of time as opportunities for emergence. As Stuart Kauffman has urged in his *Reinventing the Sacred*, "Let God be our name for the creativity in the universe. Let us regard this universe, all of life and its evolution, and the evolution of human culture and the human mind with awe and wonder."[28]

Moreover, if we freely facilitate this sacred process, we continue to participate in the cooperative constant that makes complex emergence and life possible. As a species, our choice is, once again, to learn to inter-be, and to learn to do so is to understand deeply that "the sunshine is also in this sheet of paper" and that, even though the cloud disappears after it rains, the cloud is still in the rain, and the rain and cloud are in the plant after its transpiration process. We too are like the cloud and rain and the fear of our own personal deaths is just the ego perspective of the isolated self of integrity asserting its existential crisis in a universe it fails to understand.[29]

To experience the self in this interrelated manner is to become fully conscious of others, whether they are members of our own species or other species, and realize the need for living and dying compassionately. Thus compassion, or *karuṇā*, represents the pinnacle of Buddhist practice. To extend ourselves through compassion is to experience the pathos of life and death with all of the world's beings; compassion means to become an interbeing, and as this kind of being we are held together by our subjectivity in an ecumenical web among other subjectivities in the world such as animals, plants, and stones. Ontologically speaking, there is nothing in the world (or universe) beyond interdependent arising—that is, constantly changing, impermanent, self-less processes where nothing exists separately by itself without being conditioned by other things. Again, this is the idea of *śūnyatā*, or emptiness, a lesson well learned intellectually from ecology and subjectively from our associations with other people, nonhuman animals, plants, and stones. Although we are beginning to visit these perspectives scientifically, there is a deeper religious dimension that Buddhists have always seen as being a part of the spirit of the Buddha's enlightenment.

Buddha Sâkyamuni, the original Buddha, attained enlightenment because he had direct experiences of the reality of life and death. Through this direct communion he became an interbeing with all the other beings that connected him—as through a sheet of paper—to the universe at large. To achieve compassion, we must enter ourselves and empathically extend into the experience of the other. For Dōgen, the great Japanese Zen Buddhist philosopher, Buddha-nature is the experiential presence of impermanence itself. All things are impermanent, and to profoundly experience this reintegration of the self into its eternal fluid field is to find a teacher in what we would call nature, that is, the ultimate reality that includes even such nonsentient beings as stones. When Dōgen titles a chapter "Mountains and Waters as *Sutras*" (*Sansui-Kyo*) in his masterpiece the *Shōbōgenzō*, and writes in another that "the sutras are the entire universe, mountains and rivers and the great earth, plants and trees" ("*Jishō zammai*," the *samādi* of self-enlightenment), he makes even more explicit the Buddha's pronouncements on the (enlarged) nature of the self and truth, the dharma, of all that is, including the self.[30] And to bring this point home, Dōgen writes, "People outside the way [*dao* or *dō* in Japanese] regard what is not self as the self. But what buddhas call the self is the entire universe. Therefore there is never an entire universe that is not the self, with or without knowing it."[31]

An example of this universal sense of self and its potential for compassion is cast in a deceptively negative way in the *Diamond Sutra:*

> However many species of living beings there are—whether born from eggs, from the womb, from moisture, or spontaneously; whether they have form or do not have form; whether they have perceptions or do not have perceptions; or whether it cannot be said of them that they have perceptions or that they do not have perceptions, we must lead all these beings to the ultimate nirvana so that they can be liberated. And when this innumerable, immeasurable, infinite number of beings has become liberated, we do not, in truth, think that a single being has been liberated.[32]

We suggest that the meaning here is that all must be liberated inclusively; owing to the connectivity of all things, no entity can be liberated individually. The liberation spoken about in this sutra refers to the freedom to become what we are in conjunction with all others—this is the free will of the fractal self. There is nothing particularly mystical or transcendent about this realization. This lack of mysticism and transcendent intent has led some Buddhist thinkers such as Nāgārjuna (c.150–250 CE), who is often referred to as the "Second Buddha" by Mahāyāna Buddhists, to assert that *nirvāṇa* is *saṁsāra* and that *saṁsāra* is *nirvāṇa*; that is, the realization of the ceaseless flow and change of the universe is to blow out the flame of desire, attachment, self-aggrandizement—*nirvāṇa* is simply interdependent arising (*pratītya samutpāda*).[33] *Saṁsāra* refers to an early concept from pre-Buddhist India. It is the "wheel of rebirth"—the recovery of the world after destruction.

Creation and Destruction: An Intimate Connection

Considering the equation of *saṁsāra* and *nirvāṇa,* Buddhist ecologists would tend to think of biotic succession through time, as a natural community self-organizes and develops its diversity and richness of form and function after a disturbance. For example, when a glacier melts, or a lava flow cools and crumbles, a forest begins to develop and eventually burgeons into a complex, coordinated ecosystem. This recovery could also be called the Shiva/Pele Principle because it leads to the rebirth of more creative complexity (also see chapter 4).

All "creation" also has a profound intimacy with its opposite of destruction or decay, all yin has a yang, and all yang a yin. The fact that everything decays was known by many before the advance of science. For example, in nature the waste products from all organisms become food for others. This process of biodegradation in nature, which wastes nothing material since all matter is ultimately recycled, provides the nutritional basis for the process to move along. One could term this process "creative destruction."

We are strongly reminded of the god Shiva of Hinduism—Shiva is the god who destroys the ego and self of integrity. Shiva is multifarious in form, sometimes with five heads that symbolize all that perishes and that is timeless. Shiva creates, but all creation is dependent on the destruction of what came before. This change, this flowing of all things, is Shiva's dance, a dance of life and death, death and life, evoking ecological succession. His dance is creation, destruction, and recreation in the unfolding of the universe. Shiva flirts with the edge of chaos, much like the previous trickster gods we discussed in chapter 2. He moves between life and death, reflecting the cycle of renewal. He dances on an infant, not as an act of anger and wrath, but rather through an understanding of the close relation between the old and new, the past and future.

Pele is the *akua,* a divine being who is the embodied manifestation of the creative power of volcanoes. To some she appears as the spirit of a lava flow. One of her titles is Akua lehe ʻoi, or the sharp-lipped goddess, as she consumes even rocks and trees.[34] However, while destroying everything in its path, Pele's lava provides for the future. In time the lava will slowly be transformed into rich soil, and from the soil new plants will grow after seeds drop from birds or out of the wind to await the arrival of rain and sun. When Bashō writes that he is "the same as before . . . like that aging oak" he reenters the dance of Shiva and grows in Pele's wake; a philosopher might call this is the originary moment of the Buddhist doctrine of interdependent arising, and a scientist would affirm the interconnectedness of all things, tracing back to the Big Bang.

To a Buddhist, the true reality of all that exists has been in front of our faces everywhere and from the very beginning, as in Thich Nhat Hanh's piece of paper. Moreover, this realization has a liberating effect on the self; the self is now free of its delusional isolation, the profound and constant source of suffering for humans. But we've never been isolated at all. We are connected back to the elemental. We move through the present with a real hope for meaningful participation in future universal arrangements. As Zen Master Dōgen writes, "To study the buddha way is to study the self. To study the self is to forget the self. To forget the self is to be actualized by myriad things."[35]

What we do with our time on the planet in accord with the cooperative constant has far-reaching consequences, perhaps consequences beyond our present understandings. Writ large, the Buddhist approach to ultimate compassion might view the universe as a profoundly interconnected arena of succession, and a fractal self seeks linkage with components of that succession and thus the potential to participate constructively in the universal community. The Buddha ends the *Diamond Sutra* with this *gatha*, or poetic verse.

All composed things are like a dream,
A phantom, a drop of dew, a flash of lightning.
That is how we meditate on them,
That is how to observe them.[36]

Figure 9.2 In this photograph by Nathan Wirth, the ephemeral quality of a self forgetting its self gestures toward the dreamlike condition of the meditative state being actualized by the myriad things.

The classic Buddhist view of humanity embedded in nature recognizes the evolutionary milestone of the fractal self that has emerged in our time. We cannot know the detailed unfolding of the future, but we can sense the power and the glory of the cooperative constant out of which matter and energy have assembled the magnificent complexity of the cosmos and brought us to the unique stance we assume among life-forms. The Buddhists understand that the "composed things" that preoccupy and even obsess so many of us are ephemeral. What counts most in our lives is careful consideration of the consequences of our free will and the exercise of our evolved morality toward others, leading to the sensibilities and affinities that we develop across scale within society and nature. We will prosper in the future only to the extent that we choose to sustain the progressive edge between intimacy and integrity, the source of constructive engagement with our world: from family harmony all the way to international well-being and biospheric sustainability. If we persevere, work to become sagely in our connectivity, and seek congruence with our constructive passions, we follow the way of the fractal self. As we have seen, this path continues the trend of evolution that has fostered emergence since the beginning of the universe. Choosing to follow the path blazed by the Daoists, mapped by the Buddhists, and confirmed by complexity science and modern sociobiology leads toward a potential of higher, harmonic human expression beyond our own brief time.

Nevertheless, a caveat needs to be aired. Major pathways of natural and cultural evolution, never predictable in any detailed sense, likewise perhaps have never been smooth. Emergence has sometimes revealed a dark side, turning toward simplicity, deep chaos, and destruction. Earlier we have hinted at this phenomenon as a function of the hegemony of integrity, taken toward the extreme. Human nature has spawned horrific examples of the consequences of releasing this dark cultural energy. In our time, it threatens our lives, our future, and much of the life on our planet more than ever before. As discussed in the next chapter, it often begins with intimacy—this can even be seen in Buddhism, with the mass exodus of Rohingya Muslims from western Myanmar (a predominantly Buddhist country) and Buddhist Bhutan's ethnic cleansing of the Hindu Lhotsampa minority of Nepalese origin. There has also been a resurgence of Sinhala-Buddhist nationalism in Sri Lanka against the Tamils.[37] In these rather rare cases, for Buddhists, even the most cherished compassionate precepts can be overridden by deep-seated enmities against the other. In the next chapter, we focus on how the perversion of the gifts of the fractal self on the social stage can bring dire consequences.

PART IV

The Fractal Self at Large

Anti-sage
From Cult to Empire

The illegal we do immediately. The unconstitutional takes a little longer.

—Henry Kissinger, Kissinger Cables

I am your voice . . . No one knows the system better than me, which is why I alone can fix it.

—Donald J. Trump, accepting his nomination for president by the Republican national convention

Our readers are aware by now that we define sagely behavior as benign, yet powerful, seeking toward cooperation in the world in ways that are positive, progressive, nurturing, and constructive. Sages, as we have seen, develop intimate relationships with their affinitive systems. They are naturally mutualistic, authoritative (but never authoritarian) in their butterfly-like approach to the world.

How then do we account for people who have been gifted with or have assiduously developed powers of rapport or charisma, achieving notable fractal congruence in the social, political, or economic life of institutions or communities, yet who turn those gifts and powers toward rampantly egotistical ends? This phenomenon over a wide range of scale can elevate those who become destructive or aggrandizing, to the ultimate detriment of society. History is replete with such figures.

All for One—One for Oneself

We suggest that in human nature there is a strong attraction to assertive, self-serving people who initially appear to be just the opposite. In some way, all of them may be caricatured as "confidence men." On any scale, they are consummate hucksters, demagogues, purveyors of snake-oil remedies. They may be perceived as extremely intelligent, powerful, righteous, and patriotic. Numerous followers gravitate to the kind of socially-fractally-adept individual that we call an *anti-sage*. The intuitive ways of such personages with their followers and their attainment of fractal congruence in their spheres of activity are corrupt.

Anti-sages seem to emerge in stages or with a certain deliberation; they rarely, if ever, explode on the scene. Thus we are not referring to simple despots who seize power, as in a third-world military coup. Initially, at least, anti-sagely individuals beguile rather than dictate. Typically nonsympathetic persons, they then exploit others rather than nurture them. They may or may not recognize their flawed relationship to the world around them. Instead of reacting in a self-correcting manner, they pursue aggrandizement of self to ultimate detriment or destruction. Corporation heads that pursue power and espouse greed, substitute corruption for leadership, and diminish or destroy their companies are among the most common such figures in higher levels of society today.

Anti-sages ultimately abuse the gift of intimacy or studied adroitness in systems such as business, education, politics, the military, or organized religion. Staying in power or enriching themselves motivates virtually every move they make. But with this self-serving perspective they turn away from any intrinsic connectivity or inherent leadership that encourages sustainable cooperative merit (let alone any contribution to long-lasting emergence) within their particular system. We might ask, for example, what socially valuable or progressive-constructive advances accrued to the world out of the careers of Hitler, Pol Pot, Saddam Hussein, Robert Mugabe, or most multibillionaire Wall Street bankers, the directors of Enron Corporation, et al.[1]

Historically, in the evolution of a particular system, the authoritarian posture of noncongruence inevitably precipitates the anti-sage from that system. Such a person may begin to engage with his or her social system in fractal harmony, but the relationship always breaks at some point. In some cases a critical state is reached at which the butterfly of the anti-sage *willfully* abandons any resonance with the flow of the world, and the evolving anti-sagely patterns and functions begin to fall apart. Cosmologists such as Stephen Hawking, tracing subatomic particles, refer to their trajectories as world-lines. Like the careers of anti-sages, such entities, as they race into their futures often collide with irresistible barriers and self-destruct. In both cloud chambers and large social systems, such disintegration may be catastrophic. Even the end-throes of small-scale anti-sagely movements such as the Guyana People's Temple, led to its doom by the self-styled prophet Jim Jones, can be distressingly memorable. The fate of Jones' followers has been immortalized, if not trivialized, by the endurance of the metaphor, "drinking the Kool-Aid," to express willing self-effacement or submission of followers of a failing leader in some social predicament.[2]

Masters of Deception, Disciples of Murder

Cult leaders may assume a religious mantle to mask their drive toward personal aggrandizement, and initially some of their work may resonate with positive action. Jim Jones combined religion with social work that initially impressed influential supporters such as California governor Jerry Brown and First Lady Rosalynn Carter.

Ultimately such cult leaders turn paranoid and fanatic. Cults such as the People's Temple may be essentially self-destructive and merely implode at the end, albeit tragically. Some, however, are dangerous to the greater world as their evolution proceeds to a terroristic stance directed against outsiders who become enemies of the self-identifying group and are condemned by its anti-sagely leader(s). Robert Lifton, in his analysis of the fanatic Japanese religious cult *Aum Shinrikyo,* led by Shoko Asahara, identified a phenomenon he called "altruistic murder."[3] Lifton's term refers to lethal acts and mass destruction directed against "the other" when group members come to believe such acts will lead to a greater good and a better world and sometimes to the (spiritual) benefit of the victims themselves. To any rational observer probing the psychology of this kind of system, it is usually obvious that aggrandizement on the part of the cult leader is the primary objective.

Unlike altruistic murder, with its twisted goal of saving victims from themselves and delivering them to a better world, *parochial altruism* is a perhaps ancient psychosocial phenomenon in which actors sacrifice themselves for an in-group as they seek to destroy an out-group that is considered threatening. Here the focus of the act is one-sided, and victims are merely condemned as enemies, infidels, subhuman, or the equivalent, undeserving of even the perverse consideration of spiritual benefit. In a recent study of suicide bombers by Ginges et al.,[4] research revealed that attendance at worship services—with strong bonding among congregants mediated by fanatic religious beliefs—correlated strongly with their subsequent acts of terror.

Lifton and others suggest that members of terroristic groups acquire the capacity of "psychic numbing." They abandon empathy and spontaneous intimacy with the other, as they dehumanize their often random victims, and they adopt the extreme integrity stance of their anti-sage, who preaches the necessity of their mission in the world. On the largest scale of systems that justify themselves in terms of defense of integrity or pursuit of an ideal, or purified, world are national military forces, which must tread a fine line in focusing on aggression against the *military* other in pursuing objectives, assigned by their leadership, aimed at defending their homeland or "improving" the world (as defined by their leadership).

Robert Lifton has referred to a person who follows an anti-sage, joins a cult, and thus develops an extremely narrow sense of intimacy as a "fundamentalist self." In his 1993 book, *The Protean Self,* Lifton posits a general duality of human personality. On the upside are people he calls "protean selves," after the Greek shamanic wizard Proteus who could change his form, appearing as an animal or even a natural phenomenon such as fire. Lifton envisioned a protean self as an emerging phenomenon in the modern world of multiple options and opportunities in life. During their lives, these people develop multiple interests and even careers. They shift with the times and respond to changing cultural themes, even developing notably different personalities along the way. Lifton argued that such a flexible psychological bent shapes a healthy resiliency, creativity, and satisfaction in the often-chaotic milieu of modern life.[5] On the surface, a protean self may overlap with part of our view of the fractal self but, in its "shape-shifting" progression through life, is not necessarily melding deeply in proactive, nurturing, and catalytic ways with its affinitive system(s) in nature or society. As defined, a protean self does not come to be poised, butterfly-like, for sagely development.

However, in Lifton's view, the polar opposite of the protean personality is the fundamentalist self. With this, Lifton glimpsed the looming power of the anti-sage, although the vast majority of fundamentalist selves are followers. Such individuals are looking for simplicity of thought and conviction and see the world in black and white terms (strict self vs. other), with a lack of curiosity, nuance, or willingness to grapple with complexity and uncertainty in the connected systems of the world. Many turn to religion with an unshakeable belief in mystical revelation. At their very worst, such individuals join others in the service of apocalyptic violence. They enter the ranks of Aum Shinrikyo, the Minutemen, Al Qaeda, ISIL, and their equivalents small and large. Some fundamentalist selves also seem to have a retro-vision in mind or a strong nostalgia for the past, which is identified as simpler, better, or offering a purity of lifestyle. And those prepared to follow anti-sagely directives to extreme efforts toward mass destruction also often have political and/or religious motives. In an interview with Bill Moyers just after the September 2001 attacks by Al Qaeda on the United States, Lifton suggests,

> and this kind of apocalyptic violence, the impulse is essentially religious, however dark. But there's always a combination of the religious and the political. So Asahara had political and military goals, as does bin Laden, but there must be that ultimate religious vision if you're to have this kind of extreme mass murder.[6]

However, religion in its classic guise was not an essential factor for Hitler, and not at all an issue for Pol Pot. But those figures, plus Asahara, bin Laden, and others of their ilk, invariably assume to themselves a messianic aura. When followers begin to see their anti-sage in this light, a dangerous amplification of their warped social system may ensue.

Fortunately, in many other cases, the career of an anti-sage may eventually enter a zone analogous to one in the story of "the little boy that cried wolf." Self-aggrandizement begins to falter as such individuals run out of credibility. Their system, or their constituency, slips toward stasis. An edifice of trust or confidence that has been built out of lies and distortions eventually implodes, and the individual at the apex of such a system falls into obscurity or disgrace. McCarthyism in America's turbulent political arena of the mid-twentieth century provides a vivid example. Like a cult leader, Senator McCarthy attracted slavish followers, many of them political reporters hungry for conspiratorial stories. His witch hunts were nurtured by fear that might have continued to amplify but for courageous, sagely opposition, for example by individuals such as Edward R. Murrow, Fred Friendly, and a few of their media associates of the early 1950s.[7]

In recent years, noisy and often noisome, low-echelon anti-sages have prolif-erated in popular media such as cable news and talk radio, and cults of Lim-baugh, or Beck, or Hannity echo outlandish themes and theories. The Murrows and Friendlys of our generation have lost traction. But history is not on the side of the gas-bag anti-sages of the airways, and obscurity is the fate in store for such media hucksters, purveyors of nonsense and fear, who pander to an uncritical audience, enriching themselves as they posture and cavort through the galaxies of the gullible.

Perversely, some anti-sages, after they topple from positions of leadership in government or business, having often crippled or crashed their institutions, are rewarded for their dysfunctional roles—touted by media, awarded lavish sever-ance packages, employed as consultants. This is notably the situation in the cor-porate world of the United States, suggesting a disturbing flaw in the fractal congruence of the nation's economic life vis-à-vis America's standing in the world. Such outcomes are also non-Darwinian, perhaps abetting the rise of incompetence as a national trait.

Anti-sagely America: From Admired to Mired

Careers of some US presidents, especially in recent decades, have followed distressing trajectories—leading the nation on a generally destructive path of disunity and loss of confidence. Paradoxically, such leaders may be largely passive, even congenial, figures, swayed by more-aggressive personalities in their administrations. In these cases the true anti-sage is usually in the background, but with aggrandized power, as a "senior government official," whose destructive influence on the country and the world often proceeds out of "undisclosed loca-tions." The Nixon, Reagan, and G. W. Bush administrations provide compelling examples in recent history.[8] As is common in American anti-sagely careers, such individuals as Henry Kissinger, Oliver North, and Richard Cheney—with their extreme righteousness and archaic sense of manifest destiny that is so maladap-tive in the interconnected modern world—often mislead the administrations they serve. And, nearly always, they surround themselves with subordinates of similar aggrandizing bent, some with an even more intensely expressed rejec-tion of cooperative models of the world. The closed-minded and fear-mongering attitudes of such an ideologically driven ruling clique, composed of fundamen-talist selves operating in high echelons, reinforce our nation's collective anxi-eties and too often discourage engaging with many other countries in all but an aggressively competitive or antagonistic manner. Their administrations seem likely to be remembered mainly for tragically trumped-up, unnecessary, and un-wanted wars; for negative impacts on America's international standing; and for lapses in national achievement and prominence, for example, in the sciences.

The worst effects of such anti-sagely leadership have emerged in America's re-cent foreign wars. Parochial altruism has always been a hallmark of militarism, and extremely focused examples come to mind, such as the Japanese kamikaze pilots in World War II. Beginning in Vietnam, perhaps, the degree of psychic numbing on the part of the civilian leadership of the United States rose sharply. Memory seemed to dim, amid specious legal attempts to blunt the civilized conviction, since World War II and the development of the Geneva Conventions, that directed killing of noncombatants and targeting civilian infrastructure is indefensible in "enlightened" military codification of permissible action. Of course in war zones, collateral damage (a euphemism invented by the US military) happens all the time, and berserker killings are tragically common and largely covered up, but most apologists for sovereign military actions claim they do not aim to kill the general populace of their foes. Yet, despite frequent lip service to this principle, horrendous deliberate and organized attacks across a wide scale of operations aimed at civilian debilitation and mortality were mounted by American military forces in the recent wars in Iraq and Afghanistan.

In one of the most insidious episodes of the first American war against Iraq (Gulf War) in 1991, Pentagon planners deliberately targeted the general utilities infrastructure of the country. Chalmers Johnson, in his meticulously detailed indictment of conscious infliction of civilian casualties by the US military, documents facts that rarely reached the American public.[9] During forty-three days of bombing, the United States destroyed eighteen out of twenty of Iraq's electrical generating stations and most of the water delivery and sanitary systems serving major cities. During that brief war, US Defense Department documents that were later declassified revealed that analysts anticipated the destruction of such systems would trigger epidemics of disease and mass mortality of children. In particular, these documents show there was the realization of the necessity for continuous chlorination of Iraqi water supplies to maintain public health against often-fatal infections such as cholera, hepatitis, and typhoid. Johnson continues, "Later documents state that the sanctions imposed after the war explicitly embargoed the importation of chlorine in order to prevent the purification of drinking water." Eventually, analyses relying on dozens of studies zeroed in on an estimate that, by 2001, approximately 350,000 Iraqi children had died as a result of the sanctions imposed on the import of such vital supplies. Johnson cites an observer who lived in the midst of this tragedy:

> When Denis Halliday, the United Nations coordinator in Iraq, resigned in 1998 to protest the results of the sanctions, he condemned them as "a deliberate policy to destroy the people of Iraq" and called their implementation "genocide." Given that the United States had starved the Iraqis for over a decade and caused the deaths of several hundred thousand of their children, one wonders why former deputy secretary of defense Paul Wolfowitz and others believed American invading forces would be welcomed as liberators.[10]

In 2003, the G. W. Bush administration launched the huge expansion and intensification of American foreign policy disasters in Iraq and Afghanistan, but also became synonymous with debacles at home. In one of its most devastating domestic failures, Bush-Cheney fostered deregulation of high-stakes gambling by the most powerful financial conglomerates in the world, with losses of trillions of dollars by all but those same favored players. At this time the administration acquiesced in a grossly inflating real estate bubble in the United States that was unrestrained by any semblance of sane regulation of mortgage-lending institutions. An oblivious, mostly self-and-lobbyist-serving Congress had initiated much of the irrational housing policy that the administration endorsed.

Also, with studied disregard of compelling needs for environmental regulation, Bush appointees, supported by powerful US senators and congressional leaders, notoriously denied the best scientific understanding of the burgeoning planetary climate crisis, and even attempted to sabotage the work and reputation of a top federal environmental scientist.[11] In cases of rampant neglect of pollution perils and employee-safety imperatives in the fossil fuel industries (a Cheney fiefdom), G. W. Bush and company ignored (perhaps contributed to) suppression of the best engineering practices in deep-sea drilling for oil; in fracking technology, with concomitant releases of various toxins and flammable gases into public water supplies; and also in the most fundamental safety measures in the coal mining industry.[12] Arguably, this was the most anti-sagely administration in American history.

The Obama presidency, beginning in 2008, appeared bent on healing the worst wounds of its predecessors' abusive-aggressive imperialism, but institutionalized its own dark side with its extreme international forays in Orwellian electronic surveillance and extrajudicial assassinations (with lethal collateral damage) from the skies above countries such as Pakistan and Yemen. And domestically, as it worked to bring partial recovery of one of the worst economic declines in American history, the administration all but ignored the criminality and aided and abetted the ethical abuses of the billionaire plutocrats who continue their financial and political manipulations. These are antidemocratic, anti-sagely stances in government that diminish hope for change in the United States and the world.[13] The continuing focus by the US Congress on partisanship and extreme ideology in lieu of public service is similarly foreboding (see figure 10.1).[14]

The Jesus Paradox

It is easy to assign anti-sagely status to such figures as Napoleon, Lenin, and Hitler, and to trace the known destructive and self-destructive histories of the empires built by such aggrandizing individuals. However, there are examples in which legacies of extremely mutualistic personages—who have followed true sagely paths of intimacy, compassion, and humble but profoundly emergent leadership through service that amplified in enormously influential ways—were crippled by fatal association or identification with contradictory social systems or religious beliefs. Perhaps the *personage nonpareil* of this perverse phenomenon is Jesus Christ.

The mainstream view[15] of Jesus' celebrated works (whether they were truly historical is not important here) was the essence of sagely behavior. He attracted followers effortlessly; he ministered to people's physical and spiritual needs in

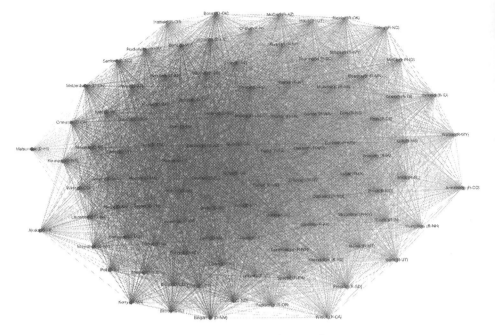

101st Congress, 1990 session

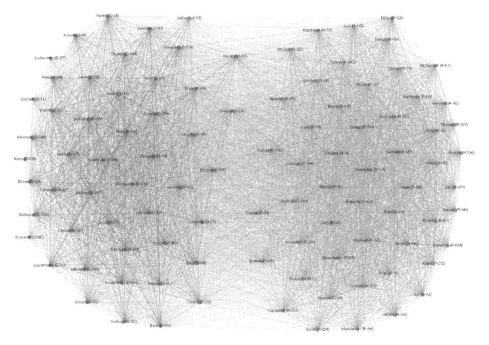

108th Congress, 2003 session

Figure 10.1

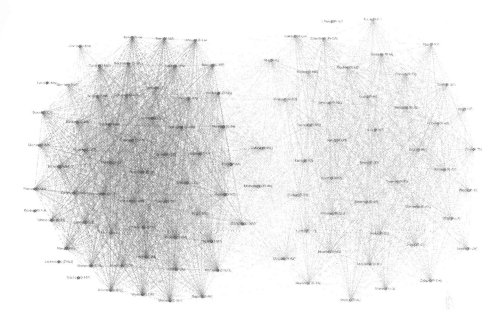

113th Congress, 2013 session

Figure 10.1 Disjunction in American government intensifying over two decades. This graph of degrees of cooperation among US senators is based on a statistical model from the physics of attraction versus repulsion over many iterations. In the graph, Democrats generally cluster to the viewer's left, and Republicans to the right.

Each crossbar represents a sitting member of the senate (use a magnifying glass to find names of the senators). The fine lines connecting particular individuals indicate that their votes have coincided at least one hundred times during the specified congressional session. Algorithmic treatment results in each senator's placement in the chart, depending on his or her cooperation in voting with all others. Thus the most bipartisan members are weighted so that they are closer to the center. The three images reflect the progressive trend of polarization in the original full set of graphs, depicting outcomes of Senate sessions from 1989 to 2013.

Renzo Lucioni, as a Harvard undergraduate in 2014, programmed the data and generated the original graphics that soon evoked similes such as an amoeba dividing and a diseased brain with its two hemispheres losing neural contact. *Source:* http://imgur.com/a/Wmoex#0. Used with permission.

a completely giving and loving manner. He worked and celebrated with them as an equal even as they acknowledged him to be an inspiring teacher and healer. The great majority of people in the countryside Jesus traveled related wholeheartedly and fractally with him. He was intimate in the holistic sense we are using in this book. Not only was Jesus easy to approach, people automatically placed trust in him. The same would perhaps be true in today's world (although one wonders how he would appear, travel, dress, and so forth in our time). Environmentalists, pacifists, and the like today evoke rueful humor imagining this: What car would Jesus drive? What kind of gun would Jesus own? As we read in the gospels, the only people who reacted negatively to him were the integrity-minded minority—the aggrandizers and power-brokers of Roman-controlled Palestine. Who might be their counterparts today? Would Jesus, appearing de novo in our time, be invited to attend Washington "prayer breakfasts?" How comfortable would he be among those nominal Christians of the political and economic elite in the United States, so eager to grasp the power of Christianity while crucifying its true spirit and mission in the world? Who would he choose as contemporary "apostles?"

In Jesus' time the ruling class feared his power of potential leadership—a power emergent through intimate association with the common people—and believed that it threatened their authoritarian systems: political stability in the case of the Romans and, more strongly, religious hegemony on the part of the conservative Jewish establishment. Exactly the same fear of the true Christian message has prevailed among the rich and powerful ever since.

The point is this: Jesus' mission was crippled from the start. Had the figure of Jesus somehow come to supplant the Old Testament angry, authoritarian, patriarch Yahweh, the world might have turned out much differently. But the punishing god, "Our Father," demanding of bloody sacrifice (including that of his own Son, whose putative nature made it partly human sacrifice) retained (or perhaps repossessed very quickly) the throne of dominance in the Christian era. The intimate message of Jesus was scattered in quiet backwaters and city slums of the world, among the same sorts of poor and powerless people he had inspired in his own time, but that message has been openly ignored, blatantly subverted, and even violently suppressed for many centuries in the great and small centers of power and wealth.

The ministering personage of Jesus that comes through the gospels and various accounts of others, such as his younger biological brother James the Just, was largely forsaken by the subsequent movement of Christianity. The advent of Pauline theology, with its emphasis on eternal reward, resurrection, and transcendence, along with its derision of the "Jerusalem-based triumvirate of James, Peter, and John"[16] effectively marginalized the intimacy Jesus had for the unfortunate and downtrodden. Paul of Tarsus' influence on Christianity and its spread is incontrovertible, but from our perspective, and following Reza Aslan's interpretation (see note 16) he comes across in his career as an aggrandizer with an anti-sagely attitude and influence that moves Christianity away from its original mission of community and compassion in its ministry. In 1 Corinthians 11:1, Paul even goes so far as to say, "Be imitators of *me*, as I am of Christ" (Aslan's emphasis). In effect Paul says, "From now on just follow me; don't look back to the original."

Born as a Roman citizen to a wealthy family, Paul's personality strongly resembles a self of integrity as opposed to the intimate self of Jesus. With and after

Paul, Christianity would become more focused on faith rather than on deeds and works, and this emphasis would later promote coercive conversion missions worldwide. In fact, Paul is principally responsible for the making of Christianity as we know it today. Combine this with the creation of the Nicene Creed—which amounts to fourth-century Emperor Constantine's order for the Roman takeover of Christianity—the soul of Christianity was overwhelmed.[17]

Thus the perversion of the Christian social gospel can be laid at the refusal of the powerful and wealthy to embrace humility, turn away from accumulating obscene riches, turn that other cheek, and especially turn from the un-ecumenical altar of God the Father, blazing with righteous rejection of non-Christian others. For centuries, Christian thought justified violent sacrifice of life in holy wars and confirmed an authoritarian, tyrannical hegemony. The true values of Jesus—openness and caring for the other, meekness, selfless leadership, humility, and all the rest—have most often gotten short shrift anywhere around the massacres, crusades, inquisitions, and colonial atrocities that "followers of Jesus" subsequently perpetrated across the globe.

Most Religions Fall from Grace

In religions of creed such as Christianity and Islam, it's easy to see how such abuses can occur. By their nature, they have the potential to evolve into all-or-nothing, strong-integrity-minded belief systems that are embellished along the way by authoritarian amendments dictated by the likes of high priests, popes, or ayatollahs to serve the agendas of the sects they control. Through time, creeds like Christianity or Islam become ultrapowerful memes that rule the lives of millions. Often, they mutate into variant ideologies of authoritarian rule across scale, from village congregations controlled by a local mullah to nation-states of the Renaissance popes and modern Iran. In any such authoritarian religion, reformers and movements that attempt to restore core beliefs—reintroducing divine attributes of intimacy, cooperation, and a participatory ethos—are met with brutal suppression.

Judaism has been differentiated from its related faiths as based more in tradition than creed. By this basis, one is born into Judaism if one has a Jewish mother, and Judaism doesn't actively seek converts to enlarge its power and authority in the world. This is not to say, however, that Judaism has been exempt from episodes of aggression invoking the power and authority of God. Such episodes became especially evident after Moses died in Sinai and Joshua led the Israelites into Palestine. As reported in scripture, God directed Joshua to eradicate pagan tribes then living in the "promised land."[18] Despite sanctioning ancient outbreaks of horrific slaughter of others, Judaism, as a religion of tradition, has been less susceptible through its later history to the more extreme acts of fanaticism that we find in both Christianity and Islam.

Buddhism itself is sometimes thought of as a religion of creed, with its rather benign, mutualistic, and simplified precepts in the Eightfold Path (see chapter 9); however, without intervention by the authoritarian God of Integrity, without the fear that drives much of Christianity and Islam, and without the institutional uncomfortableness vis-à-vis the other, Buddhism emphasizes a humanistic and naturalistic perspective of being. Buddhism has quietly assumed a global presence, albeit on a smaller scale than Christianity or Islam, which more actively seek converts. Daoism, however, and sadly enough, barely hangs on in China and is often used by the government to promote tourism. Daoist temples now comprise only about 20 percent actual Daoist practitioners, with the remainder working at the temples pretending to be Daoists for tourists.[19] Even religions of intimacy can be co-opted for less-than-admirable reasons, such as making monetary profits at the expense of honesty and authenticity, and for authoritarian gain in power and profit. In earlier chapters, we identified the urgent need to rediscover the tradition of Daoist sages, as adept players in the myriad fields of the world. Out of the same urgency, the Buddhist quest for compassion, the inclusive valuation of nature as imbued with humanity, assumes a high profile as an ethos for surviving the most pressing problems of our times. Where they resist manipulation, Buddhism and Daoism entrain models of hope; their nonauthoritarian, catalytic, and cooperative ways may well be essential to our sustainability and even survival. However, it is clear that the human condition falls far short of nirvana at this critical, dangerous phase of our cultural evolution, and that both sagely action and deep compassion are vitally needed in our near future.

Pre-dating Daoism and Buddhism in China was the Confucian religious mission, which focused more on the sanctity of human social relationships. As we discussed in chapter 8, early Confucian ideas are congruent with and have the potential for nurturing the development of a fractal self. An individual self seeks through guidance, not authoritarian commandments, supportive and cooperative roles in society by working out a path in many overlapping relational connections. As such, Confucianism is distinctive. As indicated by Henry Rosemont, "Alone among the world's religions, classical Confucianism has no monks, anchorites, or hermits in the tradition. *Only* by doing what is appropriate in your manifold relational roles as child, parent, neighbor, sibling, friend, student, spouse, and others, can you develop spiritually, as Confucius points out the path ("human way," 人道 *ren dao*) in the *Analects*."[20] Harkening back to its "First Sage," Confucianism is a religious and philosophical tradition of intimacy wherein the self, somewhat in the manner of Lifton's protean progression,[21] is continually created and recreated by its ever-changing net of relations and roles.

Sagely leaders such as Confucius and Jesus can be misappropriated and subsequently misused in ways that would leave the masters "rolling in their graves." What has become known as Confucianism has, much like Christianity, been used over the centuries to suppress the necessary free-flowing emergence across the scale of society, which is vital for humanity's health (and integrity) and accords with our main thesis in this book. Throughout Chinese history, Confucianism has often been distorted to justify the existence of a Chinese aristocracy and a static society without upward mobility, replete with sexism, homophobia, and all the ills associated with a social structure that doesn't value dynamic change. There are little, if any, references in the *Analects* to support such claims or practices, though. The famous scholar of China, H. G. Creel, captures this sense well when he writes in his *Confucius and the Chinese Way:*

> The Han dynasty, which succeeded [the Qin Dynasty], was on the whole much more favorable to the Confucians. A number of them came into conflict, however, with its sixth emperor, Wu, who had totalitarian ambitions. He was far too clever to oppose Confucianism openly; instead he posed as its patron and subsidized it. By putting large numbers of Confucians on the government payroll, and personally manipulating the examinations by which officials were beginning to be selected, he exercised considerable influence over the development of Confucian doctrine. From about this time there dates the misuse of Confucianism to justify despotism. This was a perversion of everything that Confucius stood for, against which enlightened and courageous Confucians have never ceased to protest.[22]

Following their sage, "enlightened and courageous Confucians" were always strongly interested in and dedicated to education. Our word "Confucianism" lacks the force of the Chinese designation for followers of Confucius. The Chinese term "*Rújiā* 儒家" refers to the "Family" or "School" of Scholars (often the term is translated as "the Literati"). To the Chinese mind, the intimacy of the family was something to be extended to educating the youth for a future of authentic leadership and the intergenerational quality of sharing and passing down knowledge, wisdom, and character. But throughout history in societal life—even in educational institutions, which we address next—we encounter anti-sagely forms of behavior that distort the teachings of great sages who professed intimate cooperation and compassion; subsequently, the unsuspecting and unattuned are led down a pathway of deception.

Anti-sages in Academe

The agents of doom and dysfunction that we focus on in the above sections are naturally found in sectors of society where power may promote extreme forms of social, political, and religious behaviors that feed on self-enhancement with exploitation and despotism toward others. Many transgressions perpetuated against humanity are often less spectacular than the ones we've discussed. Some anti-sagely forces that work against a system and its healthy evolution are quieter, more insidious, and sometimes arise where we might least expect. As university professors, our concern arises especially regarding the proliferation of self-aggrandizing educational administrators and how this proliferation has led us along destructive pathways away from dynamic inquiry, intellectual accomplishment, and artistic creativity. Our cooperative ideals of learning, acquiring knowledge, appreciating beauty, and achieving wisdom have progressively become lost amid abuses of leadership by administrative anti-sages. Here in our concluding section we briefly focus on higher education, primarily on leadership at the level of baccalaureate education, where we have spent our careers.

The United States, with its enormous range of pedagogy and academic research, prides itself on having the most comprehensive system of higher education in the world. However, for years, a negative trend has developed that countermands the educational mission of institutions and the lofty ideals to which colleges and universities once aspired. Instead of leading their institutions from within, many administrative leaders seek to build their own empires, seemingly with the goal of enhancing their own authority and salaries. Administrators often inflate their salaries on scales that are grossly disproportionate with faculty compensation. Manipulating and controlling those they should serve, many college and university administrators have become a power elite. These anti-sages develop cultish cadres of vice presidents and deans who begin to support the leadership over the institution. They seek to establish authoritarian control of a system whose health and dynamic well-being was once a true home of the fractal self, sustained by openness, self-organization, and emergent opportunity.

According to Benjamin Ginsberg, in his recent book *The Fall of the Faculty: The Rise of the All-Administrative University and Why It Matters*,[23] one of the most obvious indices of administrative hegemony is salary inflation. By 2011, the average pay of a president of an American four-year public university was over $440,000, substantially exceeding the salary of the president of the United States.[24] Some presidents of private schools are paid far more lavishly, and Ginsberg suggests that, "as is often true in the corporate world, the relationship between salary and institutional performance is unclear."[25] In 2010, amid national recession, the highest-paid university president was Bob Kerrey of New York City's New School. According to *The Chronicle of Higher Education* he took home over $3,000,000 that year.[26]

Anti-sagely machinations of college and university presidents occur in a number of categories. Two of the most common are insider dealings and imperial designs. University governing boards of "trustees" or "regents"—the terms vary across the country—commonly elect as members individuals from the top management of powerful corporations. And a university president serves at the pleasure of his or her governing board. In the case of insider wheeling and dealing, Ginsberg recounts a prominent episode.

One notable example is the case of Boston University whose long-time president John Silber was supported by important trustees even when it was clear that Silber had reached an advanced stage of senescence. When Silber finally stepped down in 2003 after more than thirty years in office, the incoming president, Daniel Goldin, announced that he planned to reexamine the university's business relationships with its trustees. The board responded by rescinding its offer to Goldin and paying him $1.8 million to give up the job one day before he was scheduled to take office.[27]

Empire-building by an aggrandizing college president can be a much more expansive enterprise and, in recent years, has enlarged its scope to a global scale. Arguably the most colorful recent player at this game was New York University president John Sexton. Sexton presents a complex personal and professional combined-self, perhaps straddling a chaotic edge between sage and anti-sage. According to recent profiles,[28] he took significant interest in undergraduate students and saw education as going beyond baselines of intellectual content—residing in encounters with complexity, excitement, discovery, and human values in any discipline.

Sexton's reputed darker side involved alleged secret deals with university trustees, top-level administrators, and influential members of the faculty. For example, favored individuals received payoffs, partly in the form of university-subsidized mortgages on lavish private homes. Beginning with his accession to the presidency in 2002—Sexton was appointed by the board with no other candidates ever considered—prominent board members and a few power-brokers in the administration and faculty supported the president's Olympian plans. Beginning in 2002, Sexton led the university's massive expansion, totaling millions of square feet of new construction in New York City and, through 2014, opening satellite campuses in thirteen cities around the world,[29] the latest in Shanghai and Abu Dhabi. Sexton termed his controversial and divisive grand vision the Global Network University. He believed NYU was in the vanguard of a paradigm shift in international higher education. In the new paradigm, peripatetic students will pursue learning at up to several campuses in the chosen cities Sexton has dubbed "idea capitals." Perhaps these future NYU-global graduates will emerge as planetary protean selves that would astound the likes of Robert Lifton. But perhaps there will be a tendency to seek their fortunes beyond any mutualistic engagement with the developing societies from which many of them emerged. Sexton's jocular quip to one international student—"You're going to run Ethiopia someday, right? Isn't that the goal? If you are going to run the world, it sounds like you need to get more organized"—might hint of an authoritarian attractor in the president's mind.[30]

The Global Network University emerged from Sexton's brow with, reportedly, negligible faculty participation or even consultation. In part, such imperial vision, or grandiosity, or folly as he obtained land and financing from foreign governments and even crown princes for huge faraway construction projects, triggered a vote of no confidence by NYU faculty across a wide spectrum of the university in the spring of 2013. Many other grievances were cited, among them Sexton's inflating the administration, withholding information on university policy, crushing a graduate student labor union, impoverishing faculty, and enriching cronies such as NYU vice president Jacob Lew, who became US treasury secretary in February 2013. When Lew left NYU after cutting $70 million from the university's operating budget, he got an exit bonus of $685,000 on top of his nearly $800,000 salary.[31]

Out of vision or folly, John Sexton's legacy as an academic leader may now hover between greatness and disaster. It will reveal the sign of its slope in the near future. The former president's high-flyer, risk-taker, international-educator persona seems unprecedented in American academia. Part of the NYU home campus is on Wall Street, and one wonders darkly whether any of Sexton's vision could derive from the ultimate, anti-sagely too-big-to-fail syndrome of some of his neighboring institutions.

In a number of ways, beyond the schemes of overcompensated, imperial-minded chief executives, American universities have largely adopted the corporate model in its most hierarchical form. Virtually across the entire spectrum of American higher education, administrators have propagated like an unchecked virus. Numerous vice presidents at three or more ranks head their own divisions, departments, or programs. Beneath many of them are deans, and in the descending ranks they now have associate and assistant deans ("deanlets" and "deanlings," as Ginsberg classifies them). Commonly these lower-echelon entities are now paid at the six figures level, and they have myriad staff workers at their command. Ginsberg points out that, since the 1970s, administrators in American colleges have proliferated to greatly outnumber full-time professors, and administrative staffing has bloated in numbers by 240 percent. He concludes that "colleges obviously chose to invest in management rather than teaching and research."[32]

The faculty is traditionally a body of self-governing experts in their fields who plan and carry out the actual educational mission—they develop the curriculum, teach courses, write proposals for research funds, set up laboratories, and carry out research that commonly trains students as thinkers and skilled practitioners in a discipline. Today, in many schools, the faculty has become widely perceived as relegated to an institutional tier below the administration and subservient to it, rather than the opposite relationship as once prevailed in serving the primary mission of a university.

Under administrative dictates, many a faculty member is now beset by diversions, most of them of little use in terms of fostering quality education. Many of these diversions trace to theories of higher education by Ernest Boyer in the 1990s.[33] This is only the tip of the Boyer iceberg. Much of his work initially appeared in publications of the Carnegie Foundation for the Advancement of Teaching, of which Boyer was president from 1979 to 1995. What has largely emerged from "Boyerism" is in the category of what we and numerous colleagues consider as "edubabble," which has realized enormous influence in many American colleges and universities.

Edubabble piles up in volumes of selected findings, conclusions, procedures, projections, themes, goals, objectives, learning outcomes, assessments, and so on, which are the elephantine spoor of mainly ineffective, misguided university committees, in the service of which the faculty wastes a lot of time and whose outcomes are frequently ignored. Administrators set up often-well-funded offices or centers of educational excellence that, in their wisdom, make pronouncements on the pedagogical process, hold workshops on new teaching technology, nominate task forces to study teaching effectiveness in disciplines, invent rubrics, statistics, teaching evaluation forms, and other time-sponging devices that take effort and energy away from classroom and laboratory instruction and research by professors who value their true calling: that is, the work out of which dedication develops toward realizing the beauty, rigor and complexity of truth; knowledge grows; and students may become inspired. Not all faculty members are reluctant edubabblers, however. Confirmed in Boyerism as equivalent in academic merit to disciplinary research or exemplary teaching, many a gray, vapid report on "educational process" or "strategic learning" earns its authors high praise from their ruling administrator, and this pseudo-research counts in the calculus of advancement and salary increase. The administration holds the purse strings.

Unfortunately, faculty members at many schools are now largely specialized clerks. Judgment of classroom effectiveness now comes primarily from student evaluations of their professors' "performance," as assessed by administrators focused on student retention. Weaker faculty may play to the crowd, and the depth, intricacy, and rigor of their subject gets watered down. This world has transformed the faculty into "edutainers" and purveyors of good grades in their negotiations for good student evaluations. In the end, students are reduced to consumers—some schools we know even use this term in administrative memos—who buy their educations from the marketplace of "edubiz." We suggest this new perverse sense of entitlement of students is directly linked to the rise of the "administrative university," identified by Ginsberg. As consumers, many students now believe they are entitled to a degree; formerly they were entitled to a good education.

Imperial presidents, aggrandizing vice presidents, and deans—these are the anti-sages of academe. At worst, they come to consider themselves as larger than the institution they serve. Even on lesser scales, their grasping for power subverts the educational mission, which is naturally cooperative. In much of the contemporary system of higher education, outside of the faculty, it has become rare to find a sagely individual, a devoted facilitator and intimately engaged educator. A fractal self as a faculty member is a person we can recognize immediately. He or she is one who is enamored with ideas and with communicating them. A college teacher exciting students about and with these ideas promotes any number of butterfly effects—often over and over—during the course of a semester and beyond.[34] We hope to see a shift back again to universal values of teaching and scholarship in the service of engaging student's minds with the infinite joys of discovery in the universe of learning.

Into Indra's Net

> I have great faith in optimism as a guiding principle, if only because it
> offers us the opportunity of creating a self-fulfilling prophecy. So I
> hope we've learnt something from the most barbaric century in
> history—the 20th. I would like to see us overcome our tribal divisions
> and begin to think and act as if we were one family.
>
> Arthur C. Clarke, farewell address on his ninetieth birthday

To be a fractal self is to be a participant with others, open to the worlds of nature and culture, and to become a potential facilitator of emergence. The fractal self naturally finds its place in Indra's Net and may begin to catalyze one or more cores of developing complexities as they start to amplify. Such an embedded being in the net of universal relations is in a position to propagate world-lines that may create healthy and beneficial change in the system(s) to which she or he is drawn.

We hear about such people and come into contact with some of them virtually every day. The great majority are not well known, let alone famous. Those who do ascend into popular acclaim do not seek it, but it accrues to them naturally as they work their magic within an affinitive system. Their motivation is to facilitate. Success, even fame, may emerge effortlessly, but does not drive a fractal self, ever the butterfly whose concerns and efforts remain immersed in the structure and dynamics of the cooperative milieu. Even such powerful personalities as a Gandhi or Martin Luther King Jr. largely stay focused at a sagely level within the communities they lead.

In congruence with the shared prescriptions for sagely leadership that come from Confucianism, Daoism, and Buddhism, this chapter highlights a sagely person who has exemplified the principles and actions we've been discussing in this book. He is an unsung hero in education whose career has developed in fractal leadership.

A Butterfly in Academe

Joseph L. Overton, a relatively unknown educator, has opened a wider world of intercultural understanding to students and their professors across the United States. After years of teaching, program development, procuring major grants, and creating an Institute for East Asian Studies at an East Coast university with a stifling bureaucracy and severely dysfunctional administration, he became disillusioned and dissatisfied with their lack of support. Despite institutional obstructions, and remaining sensitive to the potential for Asian Studies in US colleges, Joe cofounded the Japan Studies Association (JSA) in 1995.

In the face of his difficulties, an opportunity soon arose to foster a greater emergence in international education focused on Asia. In 1996, Joe's butterfly took off when he attended an Asian Studies Development Program at the East West Center in Honolulu. This faculty course set the stage for amplifying his efforts for the JSA. Now, over twenty years later, the JSA continues to provide comprehensive assistance in designing and expanding relevant undergraduate curricula and promoting faculty development across the country. As JSA became an effective educational bridge to Japan, Joe was also working on a much larger vision and became the creator and catalytic coordinator of the Association of Regional Centers of Asian Studies, a network of the current twenty regional centers across the United States started by the Asian Studies Development Program. Through his new role, the centers grew to provide assistance to the program's initial faculty development outreach efforts involving China, Korea, Japan, and India and to originate and maintain mechanisms for the ongoing pedagogical evolution of the community. Joe later found opportunities to develop ancillary outreach programs to a number of other countries such as Malaysia, Brunei, and Vietnam, as well as furthering even more opportunities for faculty to develop skills and gain content for teaching about East Asia. In 2000, when a faculty position opened at Kapi'olani Community College (KCC), part of the University of Hawai'i system, Joe's wings were strongly beating with resonance in the winds of higher education connecting the United States and Asia.

Prior to his arrival at Kapi'olani, Joe had secured numerous grants from such diverse patrons as the US Department of Education Fulbright-Hays Program, the Japan-US Friendship Commission, the National Security Education Program, and the Japan Foundation. However, one benefactor soon stood out. Joe's initial efforts in international education began to metamorphose with significant funding from the Freeman Foundation,[1] eventually totaling five million dollars in support of his work. In particular, the Freeman program directors recognized Joe's articulate vision of a needed educational framework and its structural details in Asian contexts, as well as his sensitivity to opening intercultural channels of understanding. Joe's catalytic role was beginning to achieve virtuoso performance.

The Freeman Foundation funded his Language and Culture Program at Kapi'olani. In designing the program, Joe and his collaborators realized that complete immersion into a culture is the optimal learning environment for students—this was one of the earliest, if not the earliest, programs of its type and has served as a model for other schools, including some of the most prestigious universities and colleges in the United States. An entire semester of coursework is devoted to learning the language of a selected country, its customs, religion, history, politics, and so forth. After completing their intensive immersion, KCC students then take courses at a sister college in Korea, China, Japan, or Vietnam, where the courses are delivered in the host country's language. As a result of their immersion, students achieve significant language acquisition and a lasting appreciation of the culture. Such experiences are not only life transforming for students, they also develop the kind of citizens needed in our new age of globalism; as many now realize, the United States can be part of a hopeful emergence for a larger, wider system of international networking that resists control and authoritarian influence.

One of the requirements of the yearlong cultural immersion program is for returning students to make a presentation in the former host country's native tongue to (and for) the Kapi'olani community. Never missing an opportunity to involve and connect as many levels of his community as possible, Joe coordinates a luncheon where students make their presentations in a festive environment. Students from the school's culinary program prepare and serve the food at the campus' restaurant. All faculty members involved in the program attend and enjoy superb Asian ethnic cuisine prepared by the future chefs of Waikiki and celebrate their students' accomplishments. Students and faculty are acknowledged and honored for their achievements. Leon Richards, Kapi'olani's chancellor, one of Joe's principal supporters, participates. Traditionally, guests of honor included Houghton "Buck" Freeman, the head of the Freeman Foundation, and his wife Doreen and son Graeme. Before his death in 2010, Buck Freeman would witness firsthand and revel in the benefits of his family's generosity unfolding in front of him. Born in China, and having lived much of his life in some of the countries in which the students studied, Buck Freeman would often converse with and ask questions publicly and privately of the students in the language of their presentations. What could be more gratifying for the leader of a generous foundation than to see his family's philanthropic seeds sprouting before his eyes? What could be more satisfying to students than to be able to demonstrate their appreciation and return their gifts of learning to their benefactors? What could be more heartwarming for faculty members than to witness the fruits of their educational nurturing efforts?

Without a supportive environment for emergence and the opportunity for educational amplification open to someone like Joseph L. Overton, this strong model of international education would have gone unrealized. As Leon Richards once remarked, "I just let Joe do what he does. I'd be a fool not to. In the process he makes us all look good, especially the Chancellor!" Engaging others to lift an affinitive system to greater levels of fullness is the essence of sagely leadership. It's with fractal wizardry that Joe choreographs the multiple levels of students, instructors, administrators, financial benefactors, and foreign hosts. In the supportive academic milieu of Kapi'olani Community College in the middle of the Pacific Ocean, Joseph Overton has become a virtuosic educator.[2]

Joe's operating style reminds us of Ding, the sagely humble cook, half glimpsed amid the implements and cupboards of his kitchen. Joe Overton foregrounds the self of intimacy and places the self of integrity in the background. Joe lives his life in relative obscurity, and in his sagely way he prefers this to the spot-light. However, he's far from obscure to the many lives he touches through the long reach of his vision and ability to execute that vision without pushing too much, or too little. In Joe's case, we see an individual enmeshed in his system, one who realizes his context at hand and moves accordingly, with anticipated outcomes awaiting manifestation on the horizons of real lives.

Joe's work reflects a passage from Confucius' *Analects* concerning the *junzi*, or consummate person (a prototype of our fractal self), and expresses a similar notion "Consummate persons establish others in seeking to establish themselves, and promote others in seeking to get there themselves. Correlating one's conduct with what is near at hand can be said to be the way of becoming consummate."[3] What sages do often goes unnoticed, for they participate in something far greater than just themselves that may unfold and amplify well into the future. The fractal self is at the heart of "all that is happening" (*wanwu*) in the world of human and humane progress.

The Nature of Our Success

Joe Overton's story highlights a kind of success in life that is very different from the popular image exemplified by many media celebrities, star athletes, Wall Street wizards, corporate moguls, assertive contemporary politicians, self-righteous religious leaders, and the like. These individuals have acquired fame or notoriety for themselves, yet their achievements most often radiate little or nothing of hallmark potential for the future. Self-absorbed aggrandizement ultimately lapses into obscurity wherever it occurs, from the most hardscrabble village to the highest ruling chambers and palaces on the planet.

Another form of faux success arises with servile sycophantic career opportunities that have proliferated in hierarchical systems of employment such as university administrations and governmental bureaucracies. In many cases these systems trace to anti-sagely leadership at the top and, as they bloat with "human resources," become dysfunctional. Eventually, they tend toward internal tribal identity and defensiveness that may injure the well-being of the greater institutions they purport to serve. At worst they become cults. The London School of Economics anthropologist David Graeber has termed most such employment in these perverted systems "bullshit jobs." In his "On the Phenomenon of Bullshit Jobs," Graeber writes that "technology has been marshaled, if anything, to figure out ways to make us all work more. In order to achieve this, jobs have had to be created that are, effectively, pointless. Huge swathes of people . . . spend their entire working lives performing tasks they secretly believe do not really need to be performed. The moral and spiritual damage that comes from this situation is profound. It is a scar across our collective soul. Yet virtually no one talks about it."[4] As we've been suggesting in this book, our collective silence is a symptom of a far greater problem, the psychological and spiritual inclinations of which are coded in the figurative DNA of how we view who and what we are and how we relate to others.

By contrast, Joe's kind of success emerges from a way of life that opens up with varying degrees of freedom to engender some significant, perhaps even profound, difference in the world. This path of a fractal self climbs high above those celebrated figures hovering or posturing in the shallows of public awareness and the overpaid drones buzzing amid the bureaucratic bullshit. As we have seen, the difference lies in the potential of a fractal self for far-reaching constructive influence, part of humanity's *evolved* capacity and affinity for catalytic contribution and cooperative action.

Sometimes we find people who have achieved wealth, status, and power in a strongly competitive milieu who, nevertheless, largely avoid or turn from the aggrandizing road. And an early aggressive instar may metamorphose to emerge as a socially constructive butterfly. In our time, ex-software-magnate Bill Gates, together with his wife Melinda, exemplifies sagely influence in both main phases of his world-line, but with his philanthropic role he has benefited countless people who would never have had a chance in the world of his former success.

As we indicated in chapter 10, wealth and power are the chief attractors of anti-sagely individuals, who ultimately foster dysfunction that may lead to disaster in the systems that they dominate. Their trajectories are similar the world over and at different scales of operation. As Americans considering political, economic, and social systems on a national scale, we naturally think first of our country. Throughout much of its history, America has fostered a visionary and strikingly innovative culture of entrepreneurs whose contributions have led the world in many fields of endeavor, harnessing forces of nature in electrical engineering, communications technology, atomic energy, and genetic research, leading inventive waves of industrial and agricultural science, pioneering powered flight and robotic exploration of the solar system, inventing personal computing and the Internet. The list is huge. Not all of those progressive individuals emerged from unheralded, unprivileged origins, but they left the world richer in potential than it had been before they appeared on the scene.

"American exceptionalism" once may have been bruited with a ring of truth, but now "too big to fail" is the hallmark of (mainly American) financial institutions; their most recent cycle of manipulation brought the whole world close to economic disaster in 2008. By virtue of measures that gauge how much these institutions and their leaders have left their country and the world poorer in potential, they deserved to fail years ago. And today the same pirates of Wall Street have returned to their rapacious ways more powerful and threatening than ever before.

Should we designate as successful institutions such as the apparel industry, whose global trade advantage has come to mean grossly exploiting and even endangering the lives of impoverished workers far from our shores? American corporations set standards as leaders of aggressively expansive, globally aggrandizing oligarchs that sway sovereign governments from behind the scenes. In other cases, self-absorbed, posturing corporate CEOs and their high-ranking managers do nothing but manipulate money to enrich their boards of directors, their shareholders, and, not least, themselves. That kind of wealth funds power, and power in such hands breeds corruption.

However, hope remains that America, assuming it does not implode from its toxic politics, is still a society that fosters sagely success with significant potential to inspire the world as it once did in many fields of endeavor. And typically the inspiration springs from individuals ranging the spectrum—from an unsung Joe Overton to iconic figures such as Elon Musk, Craig Venter, Arianna Huffington, and Neil deGrasse Tyson, to name just a few. Such people leave the self-absorbed herd far behind and follow visions that translate into careers to advance the world's science and technology, global awareness, understanding of complexity, sense of wonder, and humanistic progress.

Creed Is Not Good

Any society whose foremost identity is governed by a "sacred" creed leaves its citizens prostrate under a cold code of "divine," dictatorial commandments. Some of them unnecessarily co-opt our natural moral endowment rooted in empathy (see chapter 6); others empower an authoritarian elite of "holy" men. And in religions that mythologize forces of supernatural evil, we are yet held enthralled, as in long-past centuries of ignorance, to the "demon-haunted world."[5]

A majority of the world's nations today pay at least lip service to freedom of worship. Some enshrine this principle in national constitutions. However, within certain countries at varying scales—typically neighborhoods, villages, sometimes larger provinces—sectarian enmity smolders; extreme religious integrity promotes hatred of the infidel "other" and sometimes triggers violence.

The expressions of social cohesion with strict exclusivity in religions of creed are similar to those of cults (see chapter 10). However, both Christianity and Islam began as popular movements with sagely founders and a sense of fostering social justice under heavenly guidance. All too easily, "rulers" at all scales, claiming divine right or inspiration, appropriated the power of those faiths that came to institutionalize lack of empathy toward outsiders. For centuries, in many Christian and Islamic jurisdictions, nonbelievers could be exploited, stripped of possessions, enslaved, exiled, or simply slaughtered with presumed divine approval. Accepting conversion was often the only choice to save one's life, but didn't necessarily remove social stigma or allow for continuity of life in anything but the meanest circumstances.

Today, religious persecutions of nonbelievers and punishable rules of "blasphemy" or strict adherence for the faithful, under divine authority dictated by imams and high priests, have receded over much of the earth. A few extreme examples remain or have recently upsurged in certain notorious pockets such as West Pakistan, parts of the Arabian Peninsula, northern Nigeria, and tiny semi-isolated polygamous Mormon communities in western North America. However, in less extreme circumstances, people still struggle with paralyzing religious intrusions into politics and societal mores in countries that identify with conservative Islam and also in one country where Christian founders prescribed the separation of church and state. Among all the western democracies, only in the United States does fundamentalist religious fervor significantly encroach upon the quality and practice of public education, art, science, and medicine, representing yet another negative vector of American exceptionalism. At worst in such countries, at varying scales, a religious-political establishment seeks to abolish secular life. A fractal self, with respect to most forms of human endeavor, has little chance to grow under such circumstances.

The main antidote in these places is liberal education, which still leaves free choice in personal religiosity. Every kindergarten class is filled with fractal selves—energetic and eager to re-vision their worlds anew with humanity's strengths: empathy, cooperation, logical thinking, truth seeking, playfulness, art, science. Computers and the technology of open global communication have become difficult to suppress, even in places where schools have become sites of indoctrination and warehousing. And, fortunately, kids will always be ahead of most adults with their adeptness at using and modifying this technology. In part that may be the hope of the world.

There are signs of liberal change at many levels in the officialdom of Christendom, including some of the most conservative institutions. With the advent of Pope Francis, Catholicism is perhaps taking more significant steps than ever toward revisions of some of its hidebound policies—condemning alternatives of consenting-adult sexual identity, denying women participation in the ministry and governance of the Church, and amassing fortunes in the Vatican while many hundreds of millions of communicants live in poverty. In a July 2014 address at the University of Molise in Italy, Pope Francis condemned sinful practices of environmental degradation around the world. His main focus was on corporate greed in such industries as timber-cutting, oil drilling, and strip mining.[6] But even in this new openness to the evolving landscape of needed change, it's rather damning that this pope, as progressive as he is, still omits the huge conflict regarding his church's stance against birth control. One of the major causes of environmental decline in many of the poor countries of the world, where that church controls the lives of hundreds of millions, is overpopulation. Exponentially growing rural communities deplete regional fish stocks and wild animals on land ("bush meat"), deforest mountainsides to expand farmland and obtain firewood, and inadvertently pollute water resources in their hardscrabble struggle to survive. The church bears some responsibility for keeping those populations, which often inhabit the last, richest places for biodiversity on the planet, chained to poverty and increasingly unsustainable lives.

In recent years, a number of Protestant denominations have opened their clerical ranks to women, and openly gay church leaders are also in the ascendancy. In repressive Islamic societies, hope lies with young people who are in touch with the greater world through electronic media. Revolution at first may be necessary, perhaps in stages, as appeared so recently hopeful in Iran and other loci from North Africa to Central Asia,[7] but the dangerous ground that follows the weakening and fall of theocracy may be seized by "unholy" men whose despotic careers are as much a setback for the cooperative constant as the leaders they replace. Israel, while not politically driven by religion, represents a dilemma in the world, perhaps one that risks extreme regional discord and international violence. The country harbors an amazing compression of societal intelligence, energy, and power (that includes nuclear weapons) at the heart of its region of dire stasis and awful tragedy. For some time it has been poised as a chrysalis, but whether the trajectory of its butterfly and its potential will be dark or bright is still uncertain.

In the best of futures, cohesive humane philosophies promoting the true freedoms of individuals ought to assume that "the other" is good-natured, in the sense of Frans de Waal's understanding of our evolution of moral behavior and Mencius' proclamation of the *instinctive* core of goodness in human nature (see chapter 6). However, the tendency fostering empathy and favoring cooperation that transcends tribalism is very finely tuned indeed, and the best strategy for the long term is revealed in the winning algorithm of Prisoner's Dilemma: tit for tat. We should also look for guideposts, on the path to the future, from the wisest sources we have realized, namely reasoned interpretations of nature and human nature, and not least from the wisdom in classical Daoist, Buddhist, and Confucian philosophies. On our way we will thrive best if we share with others our own personal self-enlightening path of intimate connectivity. It leads away from authoritarian creed, commandments, and the fear of evil supernatural forces and frees us from the myth of morality as an unnatural veneer on the surface of our baser nature.

Not Man Apart

We humans have recently started taking a test. In a sense, it is an admissions examination, perhaps appropriate in our adolescence as a globally dominant species, and its successful completion relates directly to the true meaning of our free will. Knowingly ruining our rarest of planets and drastically diminishing its life would be a performance of evil unique in the history of our species. And if we ourselves should fail in the end, the failure is deeper if, through mass destruction of species and complex realms of the biosphere, we compromise or prolong some future emergence of a new eusocial consciousness. This would stand as perhaps the ultimate crime against "the other" and, as an "original sin," against our emergent universe.

A number of ecologists and philosophers have pointed out that we are already in the midst of a new mass extinction of life on Earth. If present practices of destroying the biosphere continue, this "sixth extinction"[8] will rival the worst of such former cataclysms that have been uncovered in the geological record, and we will have caused it. However, hopeful, opposing trends are still evident. Edward O. Wilson, among others, has pointed to a powerful empathic impulse in humans vis-à-vis other forms of life that he calls biophilia.[9] He suggests that it is part of evolved human nature to seek, find affinity with, cherish, and protect other life and the biosphere at large. The iconic snapshot of Kuni springs to mind—the bonobo climbing to the top of a tree and spreading the wings of a crippled bird (see chapter 6).

Modern science has discovered vital links between human welfare and the health of the planet's biosphere. Since the mid-twentieth century, a common argument for protecting the varied ecosystems of Earth has been that some of their species, many still unknown to us, will harbor biochemical compounds having properties that will prove beneficial as new medical drugs and useful industrial commodities. This has been proven again and again in organisms living on land and in the sea, and the discoveries continue unabated. Numerous other ecosystem services have been identified, all rendered free of charge to human benefit. Conserving these systems with their self-organized, self-maintaining biodiversity intact is obviously in our self-interest, considering as well such examples as watersheds sustained by montane forests, diverse fisheries fostered by estuarine nurseries of seafood species, seaside habitation protected by coral reefs and mangrove fringes, and many more. Sustaining biodiversity is not a luxury. It ought to be a responsibility, out of our sense of compassion for nature and our place within it. And finally, it is a necessity connected directly to our survival, and we dismantle it at our peril. Smooth functioning of interlocking natural systems planet-wide depends on not removing too many rivets from "spaceship Earth."[10]

The discarded rivets in Paul and Anne Ehrlich's metaphor are the species that become extinct at human hands. Some of them may be keystones in their ecosystems and, if lost, bring about further cascading extinction or dysfunction. Until fairly recently, the focus on understanding the relationship between human well-being and a healthy biosphere has generally considered this "*micro*-scale" dissected view of biodiversity. Most studies tease out the functions of an individual species without deeply considering its reciprocal relationship with the higher-level, emergent *macro* structure in which it is embedded. There may be many more keystones at varying levels in a given ecosystem. The intimacy of Indra's Net pervades all of nature: an organism needs its community of others to thrive, and the ecosystem is diminished and may collapse as it loses species.

Around the world, many nations now have legislation protecting living resources and endangered animals and plants. But even today the legal framework largely remains incomplete in laws written to safeguard biodiversity, as they consider individual species essentially in isolation. At national and international levels, showcase laws such as the United States' Endangered Species Act of 1973 and the international Convention on Biological Diversity,[11] ratified by many nations, virtually ignore the "macro" necessity of ecosystem context for species that become the objects of legal protection.

In certain ways, we ourselves are a threatened species. The prime example, irrational skeptics aside, is climate change of our own making. This threat looms ever larger and will progressively impact humanity as well as many other species and their ecosystems. Damages and disruptions will accrue to human-related offshoots of nature such as agriculture and our coastal populations and resources everywhere. Compelling analyses are emerging in support of the idea that sustaining robust biodiversity and mitigating our various environmental impacts on the balance of nature can help *us* cope with coming challenges, including climate change, some of whose consequences are now inevitable.[12] For instance, fringing coral reefs and mangrove forests, where they remain healthy, will buy time for tropical coastal communities, continuing to protect them for a while longer from superstorms, even tsunamis, as sea level inexorably rises. On some Indian Ocean shorelines, villages that had kept intact a normal mangrove waterfront largely survived the giant December 2004 tsunami, while others that had cut down the tangle of saltwater trees were swept away.[13] Here, too, we see an example of the Buddhist idea of interdependent arising that we encountered in chapters 7 and 8.

Traditionally, conservation, applied ecology, and enlightened management of wild lands and natural waters have emphasized human stewardship of nature. The concept, however well intended, implies disjunction and otherness between humanity and nature. Seeing ourselves as separate from and also set above nature, controlling it in a judicious way, may arise from the same cultural roots that gave us the Abrahamic religions with an omnipotent, authoritarian God above all (see endnote 3 in chapter 9). Stewardship implies dominion, or dominance, albeit in a caring way. However, this stance, as in religion, has always left open the door to selfish interest and abuse by aggrandizing humans on a wide range of scale. In this context, George Woodwell, former director of the Ecosystems Center at Woods Hole, Massachusetts, once asked astutely: "Do we still believe that environment is infinitely divisible by compromise each time a new claim appears?"[14]

To replace stewardship is essential, and it has become urgent to bring *partnership* with nature into the highest profile and practice everywhere. Partnership places us back in a stronger relationship with biodiversity and within biodiversity. Revised environmental legislation will wisely acknowledge and institute humanity's reembedding with nature, going forward, as we get down to the hard work of sustaining ourselves and our planetary life-support systems to secure the future. How bright or dim our own jewel remains, together with those around us in our realm of Indra's Net, depends on our imminent choice.

Leading with Our Better Angels

The urgency of concerted, cooperative action to sustain the earth's biosphere and ourselves within it has now become undeniable for anyone who relies on reason, and scientific evidence, and thinks beyond short-term parochial, economic, or ideologically driven interests. We ourselves unwittingly launched the juggernaut of climate change decades ago, and as our awareness of it has dawned slowly, the warming of the entire planet has crossed a threshold where nothing we can do will stop its initial onslaughts, likely to last for at least the next two centuries. Many impacts on human society are roughly predictable and inevitable. To name just a few: sea level rise of one to two meters; large shifts of rainfall regimes and agricultural belts to higher latitudes; a variety of increasingly severe weather conditions, both long-term and episodic; wider ranges of tropical diseases of humans, other animals, and agricultural plants; warming and acidification of ocean surface waters ruining fisheries and coral reefs. One recent analysis further suggested a direct correlation of extreme climate conditions with violence and civic and cultural unrest.[15]

The late writer Peter Matthiessen, in both his fiction and nonfiction, frequently expressed vivid positive insights regarding people from many parts of the world that he encountered in his explorations of nature and society, especially in rural and remote regions. In much of his published work, he saw hopeful instincts embedded in human heritage.

However, in his last book, *In Paradise*, the main character, Clements Olin, goes to Auschwitz for an interfaith Zen-style retreat. Olin, with whom Matthiessen admitted identifying, divulges his true purpose in being there—to search for information about his mother's fate, which had been kept from him throughout his earlier life. In his testimony to the group, Olin, a university professor from Ohio, declares that "all nations . . . and all religions, cultures, and societies throughout history have perpetrated massacres, large and small: man has been a murderer forever. A dangerous animal tragically imbalanced by its own intelligence and predisposed to violence."[16] In an April 5, 2014, National Public Radio interview, Matthiessen elaborated a bit on this sullen view of human nature, which seems antithetical to much of his earlier vision: "The number of people killed in the past century—human beings killing each other is phenomenal. . . . How has civilization—so called—come this far and people are still designing tools to kill each other? For no other purpose than killing. Why are we doing it? Why are we doing it?"

In many of Matthiessen's earlier works, he explored and wrote about the human condition intertwined with the fate of nature in remote places around the world. His approach was to immerse himself with deep empathy and intuition in environments and cultures far from the rush and turmoil and quest for plastic satisfaction of the modern world. In many rural and traditional communities he saw the hope for human and biotic sustainability that emerges with the spirit of cooperation. The worst conflict and aggression, Matthiessen noted, came out of the developed world, seething through it and radiating in all directions. In *The Snow Leopard*,[17] his reflections on a small 1973 expedition in Nepal with the field biologist George Schaller, he initially wrote of coming north out of teeming India—from Calcutta to Varanasi and on to Kathmandu and Pokhara in Nepal, then walking beyond all roads into more harmonious rural places, where "green village compounds, set about with giant banyans and old stone pools and walls, are cropped to lawn by water buffalo and cattle; the fresh water and soft shade give them the harmony of parks. These village folk own even less than those of Pokhara, yet they are spared by their old economies from modern poverty."[18]

Matthiessen's luminous quest in search of himself and the elusive big cat followed paths taken by early itinerant Daoists and Buddhists. Ultimately hiking some 250 miles through epic mountain country, he encountered the yin and yang of deep selfishness and utter generosity. Some perilous passages along the way were negotiated by prevailing combinations of empathy, compassion, and cooperation of self and other, both interpersonal and intercultural. The Sherpa leader called Tukten emerged as a striking example of a fractal self in the expedition's microcosmic reflections of hopeful cooperation in the greater world.

On his trek, passing through rural villages, Matthiessen was especially struck by the openness and acceptance of children. At a river crossing, a boy had just caught a fish. "He runs to show me, almond eyes agleam. The children all along the way are friendly and playful, even gay; though they beg a little, they are not serious about it, as are the grim . . . children of the towns. More likely they will take your hand and walk along a little, or do a somersault, or tag and run away."[19] This outreaching spirit in children generally seems part of our heritage, the "good natured" trend in primate evolution, glimpsed by Frans de Waal (see chapter 6). But to paraphrase Matthiessen's rhetoric, in his comments above on *In Paradise*, Why do we change? Why do we change?

Perhaps a tempering of that change has been emerging, all but imperceptibly until now. A new view that has arisen from exhaustive scholarly research comes from the Harvard psychologist Steven Pinker. He concludes in his recent book, *The Better Angels of Our Nature*,[20] that violence in human affairs has diminished through history. He traces this change across scale from family strife to conflict between nation-states. As the complexity of human society evolved out of bands of hunting and gathering nomads to the regulated lives of citizens in large governed societies, Pinker finds evidence and sifts data with statistical treatments that indicate manyfold declines in rates of violent death imposed by other people. The trend appears to hold for both organized warfare and simple murder.

Pinker refers to progressive "revolutions" that have correlated with violence reduction. Beginning with "pacification," which partly mediated the escape from the Hobbesian dilemma,* the revolutions have proceeded to "civilizing," as people not only identified with their system's polity and authority but also traveled farther and conducted business abroad, then to "humanitarian," as people considered human rights in deeper contexts, including relief from abuses such as slavery. Of course these revolutionary shifts in cultural-ethical complexity have occurred on varying schedules in different parts of the world. Pinker concludes his tour of these emergent historical trends with his "rights revolution," which today concerns treatment of others on regional to global scales: minorities, women and children, homosexuals, and even animals. Nevertheless, rights typically imply a legal context and need for enforcement. With a hopeful outlook, we might imagine at least a new revolutionary stage emerging that will aim toward overcoming persistent tribalism. Such an "intimacy" revolution, inspired by classic Confucian, Buddhist, and Daoist principles, and even ecological succession,

---

* As we noted in the introduction, Hobbes proposed in 1651 the need for (inferred) authoritarian rule to reduce interpersonal and community violence in ungoverned society. Pinker views this as a stage of social evolution, and as we've argued throughout this book, such a model of societal leadership is not sustainable in our world today.

would approach a new global level of intercultural respect, international coop-eration, and biospheric sustainability.

Pinker's analysis of pacifying, humanizing, and civilizing trends in social evolution perhaps bespeaks a strengthening cooperative constant in play, but critics point to huge oscillations in levels of violence during the twentieth century, and Pinker himself does not attempt to project his thesis into the future. However, if he is right, he has bolstered our hope for the near future, at least, in which the potential for social strife and environmental crisis loom so large. His book's title comes from Abraham Lincoln's Gettysburg Address, and his "angels" are four: empathy, self-control, morality, and reason—human qual-ities that appear rooted in primate evolution (see chapter 6). They are qualities that begin to emerge in children everywhere. They become cardinal virtues of a fractal self and are ultimately nurtured and strengthened in the practice of sagely behavior.

Existential Risks

We are all children of nature who have only begun to evolve into gods. Our im-mense and growing powers extend to nearly all realms of our planet and now even stretch a little beyond. Many technological advances, chiefly in the last century, have brought human beings to this threshold. Three, in particular, now represent not-so-securely bottled genies that project the greatest uncer-tainties in our near future. The first, nuclear science, was quickly harnessed to produce weapons of mass destruction, but was secondarily developed for peace-ful applications. However, only a few of its uses, such as in medical treatments against cancer and age-dating ancient geological strata and fossils, have avoided dangerous potential consequences (as attend nuclear electrical generation, for in-stance). Above all, rational minds have come to understand that nuclear weapons are unusable. And the very plausible fears of international nuclear spasm[21] that terrified the world through several decades of the twentieth century—exchanges of hundreds or thousands of bombs that would effectively extinguish civilization—have largely subsided. Danger still lurks in fanatic minds that would trigger a rogue weapon in a major world city. With time, through erosion of perverse misguided moralities and nurturing of cultural tolerance, we may hope that this threat becomes progressively improbable.

The second comes from our ability to decode, transform, and manipulate the molecules of life. Genetic engineering is a promethean technology whose power to vitally affect us and our biosphere easily rivals that of nuclear energy. Some of these applications may be highly beneficial—in medicine, in agriculture, and in tailoring bacteria for industrial uses also in cloning animal tissue and certain animals themselves. Or instructional—for understanding life's evolutionary his-tory and constructing artificial cellular life. Some other applications may be frivolous—for instance, resurrecting extinct animals whose ecosystems are long gone. *Do no harm* must be the essential guiding principle in twenty-first-century genetics. Nightmare scenarios have been envisioned of creating novel pathogens against which we are defenseless. And with the lucrative rush to engineer and fine-tune the blueprints of our agricultural species, we must not discard the gene pools that were selected by traditional farmers over the past ten thousand years. If the wisdom of our actions in this field does not catch up to dazzling laboratory discoveries and technological prowess, we will be in a world of folly.

The contemporary acceleration of human capability to wield godlike powers engineering genomes perhaps exceeds all previous technological progress in applied biology. The CRISPR revolution that began after 2010, enabling cheap, rapid genetic alteration of any life-form on Earth, expanded worldwide in half a decade (see also chapter 3). In April 2015, a research team in China announced they had successfully manipulated the DNA of human embryos with extreme precision. At the time, they emphasized that those embryos were nonviable. Their goal, together with colleagues in many nations today, is to apply the breakthroughs in their field to gene therapies against cancer, birth defects, and so on. However, the sense of urgency over the power of CRISPR to change many aspects of life itself led to an international congress of top researchers in December 2015. Their resolution called for the strongest regulation of this technology, lest it lead to unethical and dangerous developments.[22]

The third potential existential hazard has so far been defined less crisply than nuclear terrorism and dark biotechnology. It has been envisioned as stemming from rapidly progressing research in artificial intelligence. For decades, much of the discussion of the future significance of AI has been the province of science fiction. Computer theorists, software engineers, neuroscientists, psychologists, and analytical philosophers generally admit they have not yet fully characterized "intelligence" per se. However, new developments, some of them coming from game theory, and especially from a burgeoning field of algorithmic design called deep learning, appear to be driving "machine intelligence" into rapid ascendancy in the real world.

A comprehensive review of these developments and their possible consequences appeared in the 2014 book *Superintelligence: Paths, Dangers, Strategies*,[23] by Nick Bostrom, a philosopher and logician who, in 2005, founded a think tank called the Future of Humanity Institute at Oxford University. Bostrom has attracted to the institute an ensemble of forward-looking intellectuals from a wide range of disciplines: physics, neuroscience, computer science, robotics, economics, and others. In the last decade several similar organizations around the world also have begun to probe the risks of human extinction posed by advances in "strong AI." Funding has grown rapidly. A growing list of luminaries in science and technology, among them Stephen Hawking, Martin Rees, Max Tegmark, Elon Musk, and Bill Gates, have publicly aired strong concerns that echo Bostrom's warnings.

In the mid-twentieth century, cyber visionaries including Alan Turing, Norbert Weiner, and others first warned of runaway computer capabilities. One Turing associate, I. J. Good, anticipated Bostrom's thesis. Good wrote, "An ultraintelligent machine could design even better machines. There would then be an 'intelligence explosion' and the intelligence of man would be left far behind."

He went on to suggest that we might get lucky if the machines remained benign to us.[24] Nick Bostrom finds a less optimistic outlook in primate evolution, comparing us to gorillas, which we have left far behind and on the verge of extinction.

Aside from notions that echo Hollywood plots of humanoid robots rising to become a new ruling class and lording it over humans, a kind of primitive AI revolution seems already in progress in the form of highly distributed, rapidly diversifying, and increasingly sophisticated personal and institutional computerized systems that globally change our behavior and cultures as never before. They often have an almost irresistible appeal and grow to seem essential to our lives. Some addict us as powerfully as any drug. The systems and devices that deliver their apps steadily become more powerful, store more information, and have begun to interface automatically and exchange data about us. This new world order includes vast networks of mobile phones, programmed with multiple functions; in our homes, online monitors and reporting modules within home appliances and security systems store up our schedules and preferences. We are becoming accustomed to robotic cars and trucks on the roads, drones in the air, Fitbits reporting on our bodily functions. Are we already being pulled toward some sort of vast socio-technological compulsion that begins with irresistible appeal and then becomes impossible to resist. Bostrom's main point seems to be that control of what used to be our world could devolve to dominance by AI that ultimately would manifest indifference or competition with us. Perhaps we are even now setting ourselves up for a fall if self-replicating, self-improving algorithmic minds become self-aware. Bostrom calls this possibility our greatest existential risk.

By contrast, perhaps, our foresight might create AI to represent our "better angels." That vision began in science fiction decades ago with stories such as Isaac Asimov's "The Last Question." Recently, in his novel *Aurora*, Kim Stanley Robinson presented a more mature view of what an intimacy-based symbiosis between humans and strong AI in the next few centuries might look like. Even with all of our burgeoning, device-laden distractions, the memes that rise to the top and take hold in our wide hyper-connected, still troubled world are often those of compassion. Might this herald an interdependent arising across scale, empathic world-lines converging beyond tribalism, even pervading "machine intelligence," uniting human beings with intimacy for each other and for the living Earth?

The next several decades may bring our greatest trials of all time. We believe that the human condition and experience has a good chance to evolve toward a mature future of greater wisdom, achievement, compassion, and satisfaction. Much of our support in reaching that future will come from the best of what biological nature has bequeathed us—from the hopeful, cooperative, good-natured side of our evolutionary history, and also from within the wonderfully rich and complex biosphere to which we belong. At this time of our testing, our future can be bright if our answer is that we *choose* to coexist with other selves and with nature wisely and in benign partnership.

Where *Will* We Go from Here?

In the worldview of the philosopher and cosmologist Julian Barbour,[25] time is an illusion, there is never any present moment between past and future. And hence past and future also do not exist. Everything is in the same frame, a tunnel of timeless existence. Some sectors of the picture contain us in familiar complex surroundings: galaxies, stars, planets, and life. Other realms contain next to nothing. In Barbour's universe they all exist simultaneously.

Ever since Newton, physics has grappled with the uneasy logic that our sense of time is a fallacy. Newton's universe is deterministic, in that the motions of all of its working parts, and parts of those parts, are subject to universal laws. And if those motions were to be measured or calculated to the ultimate degree, the future is totally predictable, and thus time, and the freedom that would unfold with it, are irrelevant. There can be no free will, no choice at a particular moment that will change the future.

Einstein roiled the waters that had cleared for Newton by showing that time is relative according to the observer. In the Theory of Relativity, time slows for any object or system as its velocity increases and would cease if one could reach the speed of light. Thus time is not absolute in Einstein's universe and, again, becomes essentially irrelevant, as it is in that other great revelation of twentieth-century physics known as quantum mechanics. Quantum theory greatly disturbed Einstein, who objected to the idea that subatomic particles were essentially unpredictable in their behavior and could be in more than one place at the same instant. We can only locate them in terms of their probability of occurrence.

Very recently the quantum theorist and cosmologist Lee Smolin, in a highly speculative search for new physical principles, is attempting to bring an immutable version of time back into the picture.[26] In Smolin's view, physical laws themselves, and basic parameters and settings of our universe, may evolve; at the most basic level this could confirm the primacy of time and history, and freedom's potential for the future.

Regardless of these theoretical conundrums that rise like spectres from the deepest ether of quantum uncertainty and relativistic horizon, there is a verifiable thread in our universe, emerging out of simplicity to give rise to progressively more complex states and systems. Measureable facts, evidence recorded in strata of categories physical, chemical, geological, and biological—within observable worlds of systems evolving by cause and effect, self-organizing, turning chaos into leaps of emergence—this is our lived-in science and philosophy that gives us hope of navigating a future with sagely vision, skill, and compassion. And in our world-frame, far above quantum bottom, fractal selves find freedom to live creative lives and perhaps even sometimes, on the far limb of the power curve, change the world. For most of us, the outlook in our worldlines is not as grand, but for every human being, out of our free will, there is a seminal choice that leads us toward and through a life in which we, our society, and the rest of our sustaining world hangs in the balance.

As realized in chapter 9, we are the first and only life-form that we know having the willful ability to tip the balancing act of our evolutionary heritage—to foreground the simple path of vacuous integrity that will leave us in a frozen realm with our freedom lost or diminished in degree. At the present crossing, we make our choice. Ahead of us, the continuing main road of intimacy extends far beyond our field of vision. Behind us we know that it has always led through organic ascendance. Now, our very survival may depend on whether we continue on that road. Beyond survival, our species' ultimate success depends on whether we will favor in our lives and societies the cooperative constant that has brought us to such a stunning threshold, the most significant milestone in 3.7 billion years of life's evolution: making creation, in every way, a conscious process for the first time. *Will we be up to it?* This simple question is the most profound ever posed to life on Earth.

NOTES

Introduction

1. Martin Heidegger, in his *An Introduction to Metaphysics*, considered this question as being the most fundamental issue of philosophy. See Martin Heidegger, *An Introduction to Metaphysics*, trans. Gregory Fried and Richard Polt (New Haven, CT: Yale University Press, 2014), 1–2. The question was stated long before Heidegger in an essay by Leibniz in 1714. See G. W. Leibniz, "Principles of Nature and of Grace Founded on Reason," trans. G. H. R. Parkinson and M. Morris, in *Leibniz: Philosophical Writings*, ed. G. H. R. Parkinson (London: J. M. Dent & Sons, 1714/1973), 195–204.

2. J. Gleick, *Chaos: Making a New Science* (New York: Penguin Books, 1987).

3. C. G. Langton, "Computation at the Edge of Chaos," *Physica D* 42 (1990): 12–37. Also, J. P. Crutchfield and K. Young, "Computation at the Onset of Chaos," in *Entropy, Complexity, and the Physics of Information*, ed. W. Zurek, SFI Studies in the Sciences of Complexity, vol. 8 (Reading, MA: Addison-Wesley, 1990), 223–269. Controversy over theoretical indications of richness of complexity peaking near the "edge of chaos" was first raised by M. Mitchell, P. T. Hraber, and J. P. Crutchfield, "Revisiting the Edge of Chaos: Evolving Cellular Automata to Perform Computations," *Complex Systems* 7 (1993): 89–130.

4. E. N. Lorenz, "Deterministic Nonperiodic Flow," *Journal of Atmospheric Sciences* 20 (1963): 130–141.

5. E. N. Lorenz, "Predictability: Does the Flap of a Butterfly's Wings in Brazil Set Off a Tornado in Texas?" Address, 1972: American Association for the Advancement of Science, https://www.technologyreview.com/s/422809/when-the-butterfly-effect-took-flight/

6. G. A. Held, D. H. Solina, D. T. Keane, W. J. Haag, P. M. Horn, and G. Grinstein, "Experimental Study of Critical-Mass Fluctuations in an Evolving Sandpile," *Physical Review Letters* 65 (1990): 1120–1123. Also see P. Bak and K. Chen, "Self Organized Criticality," *Scientific American* 264, no. 1 (1991): 46–53.

7. B. Mandelbrot, *The Fractal Geometry of Nature* (New York: W. H. Freeman, 1982).

8. An excellent, vividly written, journalistic account of the foundations and development of complexity theory is found in M. M. Waldrop, *Complexity: The Emerging Science at the Edge of Chaos and Order* (New York: Simon and Schuster, 1992).

9. See for example, S. Kauffman, *At Home in the Universe: The Search for Laws of Self-Organization and Complexity* (London: Oxford University Press, 1995).

10. A thought experiment made popular in the writings of Stephen J. Gould, initially appearing in S. J. Gould, *Wonderful Life: The Burgess Shale and the Nature of History* (New York: Norton, 1989).

11. S. Kauffman, *Reinventing the Sacred* (New York: Basic Books, 2008), see page 100 and chapter 8.

12. J. A. Estes and J. F. Palmisano, "Sea Otters: Their Role in Structuring Nearshore Communities," *Science* 185 (1974): 1058–1060.

13. See, for example, P. Hammerstein, ed., *Genetic and Cultural Evolution of Cooperation* (Cambridge, MA: MIT Press, 2003).

14. M. A. Nowak, with Roger Highfield, *SuperCooperators: Altruism, Evolution, and Why We Need Each Other to Succeed* (New York: Free Press, 2011). Also, M. A. Nowak, "Five Rules for the Evolution of Cooperation," *Science* 314 (2006): 1560–1563.

15. T. P. Kasulis, *Intimacy or Integrity: Philosophy and Cultural Difference* (Honolulu: University of Hawai'i Press, 2002).

16. See, for example, P. K. Dayton and R. R. Hessler, "Role of Biological Disturbance in Maintaining Diversity in the Deep Sea," *Deep-Sea Research* 19 (1972): 199–208.

17. Thomas Hobbes, *Leviathan, or the Matter, Forme, and Power of a Commonwealth, Ecclesiasticall and Civil.* One modern reprint is Thomas Hobbes and C. B. Macpherson, *Leviathan* (London: Penguin, 1988 [1651]).

18. Beginning in the 1980s, unusual bacteria and related life-forms began to be discovered up to several kilometers deep in the earth's crust, beneath the ocean floor, and in land deposits around the world. Some scientists have termed this hidden life zone the "deep hot biosphere." See, for example, K. Lysnes, T. Torsvik, I. H. Thorseth, and R. B. Pedersen, "Microbial Populations in Ocean Floor Basalt: Results from ODP Leg 187," in *Proceedings Ocean Drilling Program Results* 187 (2004): 1–27, ed. R. B. Pedersen, D. M. Christie, and D. J. Miller.

19. R. Dawkins, *The Selfish Gene* (New York: Oxford University Press, 1976); and other works, for example, S. Blackmore, *The Meme Machine* (New York: Oxford University Press, 1999).

Chapter 1: Primal Emergence

Epigraph. A. Linde, "The Self-Reproducing Inflationary Universe," *Scientific American* 271, no. 5 (1994): 48–55.

1. Cosmologists, most of whom are also physicists, have progressively fine-tuned models of the evolving universe with many points of agreement. After the 1980s, the pace of astrophysical discoveries accelerated, and new insights honed and pruned classical conclusions. Some general overviews of this progress in cosmological understanding can be found in S. Weinberg, "Life in the Universe," *Scientific American* 271, no. 4 (1994): 44–49. Also see A. Guth, *The Inflationary Universe: The Quest for a New Theory of Cosmic Origins* (Reading, MA: Perseus, 1997); Brian Greene, *The Fabric of the Cosmos: Space, Time, and the Texture of Reality* (New York: Alfred A. Knopf, 2004); and L. M. Krauss, *A Universe from Nothing: Why There Is Something Rather than Nothing* (New York: Free Press, 2013).

2. J. Cartwright, "Evidence of Antimatter Anomaly Mounts," Science NOW (AAAS, 2012), http://news.sciencemag.org/sciencenow/2012/02/evidence-for-antimatter-anomaly-.html.

3. A lucid, readable source is L. Susskind, *The Cosmic Landscape: String Theory and the Illusion of Intelligent Design* (New York: Little, Brown, 2006). Also, see Linde (1994).

4. The source for this is from Simplicius: "For those who supposed the worlds to be infinite in number, like the associates of Anaximander and Democritus and afterwards those of Epicurus, supposed them to be coming-to-be and passing away for an infinite time, with some of them always coming-to-be and passing away; and they said that motion was eternal." Quoted from G. S. Kirk and J. E. Raven, *The Presocratic Philosophers: A Critical History with a Selection of Texts* (London: Cambridge University Press, 1957), 124.

5. J. C. Gregory, *A Short History of Atomism: From Democritus to Bohr* (London: A. & C. Black, 1931), 258. Also S. Berryman, "Democritus," *The Stanford Encyclopedia of Philosophy,* ed. E. N. Zalta, 2010, http://plato.stanford.edu/archives/fall2010/entries.

6. Cicero translated Aristotle's four kinds of *aitia* into the Latin *causae,* where our word "causes" comes from. Hence we've inherited the terminology of Aristotle's Four Causes. Randall writes that "[the Greek word] *Aition* means literally the answer or response to a question; it meant in Greek what could be held 'answerable' or 'responsible' in a law court. Aristotle's four *aitia* are the four different factors 'responsible' for a process, the four 'necessary conditions' of any process, four *dioti*'s or 'reasons why,' four 'wherefores.'" See John Herman Randall Jr., *Aristotle* (New York: Columbia University Press, 1960), 123–124.

7. As Aristotle writes of the Prime Mover: "Therefore it must be of itself that the divine thought thinks (since it is the most excellent of things), and its thinking is a thinking on thinking" (295) and "It is clear then from what has been said that there is a substance [*ousia*, fundamental being] which is eternal and unmovable and separate from sensible things" (286). This is all prefaced by Aristotle's conclusion: "There must, then, be such a principle, whose very essence is actuality" (282). See Richard McKeon, *Introduction to Aristotle* (New York: Random House, 1947).

8. For a lucid account, see L. Looney, Department of Astronomy, University of Illinois at Urbana-Champaign (web pages for this astronomy class illustrate a brief history of the universe), 2011, http://eeyore.astro.illinois.edu/~lwl/classes/astro330h/spring11/Lectures/lecture5 .pdf. Also, L. M. Krauss (2013).

9. See Greene (2004).

10. Weinberg (1994).

11. W. Hu and M. White, "The Cosmic Symphony," *Scientific American* 290, no. 2 (2004): 44–53.

12. The discovery of the CMB—remnant "glow" of the Big Bang—was reported by A. A. Penzias and R. W. Wilson, "A Measurement of Excess Antenna Temperature at 4080 Mc/s," *Astrophysical Journal Letters* 142 (1965): 419–421. Also, collaborators at Princeton University interpreted the cosmological implications of the CMB. See R. H. Dicke, P. J. E. Peebles, P. J. Roll, and D. T. Wilkinson, "Cosmic Black-Body Radiation," *Astrophysical Journal Letters* 142 (1965): 414–419.

13. First detection of anisotropy (or nonhomogeneity) in the CMB was discovered by George Smoot and colleagues. A readable account is found in G. Smoot and K. Davidson, *Wrinkles in Time* (New York: William Morrow, 1994), 331. For a modern update on subsequent improved resolution of the CMB, see for example: http://map.gsfc.nasa.gov/.

14. Ahead of his time, Charles Peirce probably would not have had trouble fitting concepts such as quantum physics and chaos theory into his worldview. See, for example, http://plato .stanford.edu/entries/peirce/#anti.

15. See any current astronomy textbook.

16. See, for example, R. Panek, *The 4 Percent Universe: Dark Matter, Dark Energy, and the Race to Discover the Rest of Reality* (New York: Houghton Mifflin Harcourt, 2011), 320. Also, a short summary of the detection of the onset of faster expansion of the universe appeared in http://www.nytimes.com/2007/03/11/magazine/11dark.t.html?pagewanted=all.

17. G. Dvali, "Out of the Darkness." *Scientific American* 290, no. 2 (2004): 68–75.

18. See any recent astronomy textbook.

19. See, for example, chapter 4 in D. Prialnik, *An Introduction to the Theory of Stellar Structure and Evolution* (New York: Cambridge University Press, 2000), 263.

Chapter 2: Out of the Dreamtime

Epigraph. Friedrich Nietzsche, *The Birth of Tragedy and the Genealogy of Morals*, trans. Francis Golffing (Garden City, NY: Doubleday, 1956).

1. *Thumos* is nearly impossible to translate from the Greek. It can mean spirited, soul, heart, mind, temper, passion, anger, will, courage, and temper. *Kradie* and *etor* are a bit easier and refer to the heart, and *phrenes* to the lungs. All these words refer to more than just their physiological functions and refer to thinking, consciousness, and so forth. For a good discussion of this multiple sense of the psyche, see Bruno Snell, *The Discovery of the Mind in Greek Philosophy and Literature* (New York: Dover, 1982; Harper, 1960). For an interesting read on the relation between the Homeric sense of self and the psychological, see Julian Jaynes, *The Origin of Consciousness in the Breakdown of the Bicameral Mind* (New York: Mariner Books, 2000).

2. Owen Barfield, *Saving the Appearances: A Study in Idolatry* (New York: Harcourt Brace Jovanovich, 1957), 40.

3. See R. H. Codrington, *The Melanesians: Studies in Their Anthropology and Folklore* (London: Clarendon Press, Oxford University, 1891), 119.

4. Robert Lawlor, *Voices of the First Day: Awakening in the Aboriginal Dreamtime* (Rochester, VT: Inner Traditions International, 1991), 1.

5. For a complete study of the Navajo creation myth, see Jerrold E. Levy, *In the Beginning: The Navajo Genesis* (Berkeley: University of California Press, 1998).

6. See Martha Warren Beckwith, *Hawaiian Mythology* (Honolulu: University of Hawai'i Press, 1970; Yale University Press, 1940).

7. Fragment 64 and Fragment 66. David Jones translation, previously unpublished.

8. See Graham Parkes, *Composing the Soul: Reaches of Nietzsche's Psychology* (Chicago: University of Chicago Press, 1994), 67.

9. Friedrich Nietzsche, trans. Francis Golffing, *The Birth of Tragedy and The Genealogy of Morals* (Garden City, NY: Doubleday, 1956), 63.

10. Golffing (1956), 97.

11. E. R. Dodds, *The Greeks and the Irrational* (Berkeley: University of California Press, 1951), 140.

12. Dodds (1951), 140.

13. Mircea Eliade, *Shamanism: Archaic Techniques of Ecstasy* (Princeton: Princeton University Press, 1974; Bollingen Series 2004), 182.

14. This movement from the many to the one can also be seen in early Vedic literature. In the *Ṛg Veda* there is "the transition from naturalistic polytheism to monotheism and then to the philosophical monism, which constitutes the main philosophical doctrine of the Veda later to be carried over into the Upaniṣads and eventually into the most highly developed system of Indian thought, the Vedānta" (16). As this transition progressively takes place in the *Vedas*, we find the statement "Let all the others die away" repeated again and again as if a mantra. From Sarvepalli Radhakrishan and Charles A. Moore, eds., *Source Book in Indian Philosophy* (Princeton: Princeton University Press, 1957).

15. Norman O. Brown, *Theogony-Hesiod* (Indianapolis: Liberal Arts Press, 1968), 56.

16. Brown (1968), 57.

17. Ibid., 58.

18. Ibid.

19. Ibid., 59. Italics in the original.

20. See Friedrich Nietzsche, trans. Walter Kaufmann, *Beyond Good and Evil: Prelude to a Philosophy of the Future* (New York: Vintage Books, 1966), 3. Nietzsche's full quote is even more revealing: "But the fight against Plato or, to speak more clearly and for 'the people,' the fight against the Christian-ecclesiastical pressure of millennia—for Christianity is Platonism for 'the people'—has created in Europe a magnificent tension of the spirit the like of which had never existed on earth: with so tense a bow we can now shoot for the most distant goals."

21. Alfred North Whitehead, edited by David Ray Griffin and Donald W. Sherburne, *Process and Reality* (New York: Free Press, 1978), 39.

Chapter 3: The Quickening of Chemistry

Epigraph. Albert Claude, "The Coming Age of the Cell," Nobel lecture, December 12, 1974, https://www.nobelprize.org/nobel_prizes/medicine/laureates/1974/claude-lecture.html.

1. Archaean catalytic clays have been suggested by many researchers as having a strong potential for organizing early biomolecular synthesis. See, for example, Alain Meunier, Sabine Petit, Charles S. Cockell, Abderrazzak El Albani, and Daniel Beaufort, "The Fe-Rich Clay Mi-

crosystems in Basalt-Komatiite Lavas: Importance of Fe-Smectites for Pre-Biotic Molecule Catalysis during the Hadean Eon," *Origins of Life and Evolution of Biospheres* 40, no. 3 (2010): 253–272.

2. F. Hoyle, *The Intelligent Universe* (London: Michael Joseph, 1983). Also, I. Musgrave, "Lies, Damned Lies, Statistics, and Probability of Abiogenesis Calculations," 1998, http://www.talkorigins.org/faqs/abioprob/abioprob.html.

3. M. J. Behe, *Darwin's Black Box: The Biochemical Challenge to Evolution,* 2nd ed. (New York: Free Press, 2006).

4. For example cilia shifted from being mechanical cellular organelles to the light-sensing rods and cones in vertebrate eyes, and hinge bones in reptilian jaws became the articulated middle-ear ossicles that transmit sound from the eardrums of mammals. Examples abound of the emergence of novel functionality in the evolution of biological structure. See any college textbook in general biology.

5. See, for example, W. L. Brown Jr. and E. O. Wilson, "Character Displacement," *Systematic Zoology* 7 (1956): 49–64. Also, S. Wright, "Character Change, Speciation, and the Higher Taxa," *Evolution* 36, no. 3 (1982): 427–443. Also, S. Kauffman, *At Home in the Universe: The Search for the Laws of Self-Organization and Complexity* (New York: Oxford University Press, 1995) (see 149–189 for extended discussion and recent insights). Also, S. Sahney, M. J. Benton, P. A. Ferry, "Links between Global Taxonomic Diversity, Ecological Diversity and the Expansion of Vertebrates on Land," *Biology Letters* 6, no. 4 (2010): 544–547.

6. Ribosomes, intricate scaffold-like structures of cooperating RNA and protein molecules, are tiny assembly units for translating the genetic code from its nucleic acid base into the sequenced amino acid chains that become the working proteins for all cellular functions. See a good biology text.

7. The renowned thirteenth-century Japanese Zen master and philosopher Dōgen expresses this sense of nested interconnectivity in his Bendō-wa essay of the Shōbōgenzō (*Treasury of the True Dharma Eye*): "Because earth, grass, trees, walls, tiles, and pebbles all engage in Buddha activity, those who receive the benefit of wind and water caused by them are inconceivably helped by the buddha's guidance, splendid and unthinkable, and awaken intimately to themselves." See Kazuaki Tanahashi, *Moon in a Dewdrop: Writings of Zen Master Dōgen* (New York: North Point Press, 1985), 146. For an insightful interpretation of the importance of stones in Chinese and Japanese cultures, see François Berthier, *The Japanese Dry Landscape Garden: Reading Zen in Rocks,* translated and with a philosophical essay by Graham Parkes (Chicago: University of Chicago Press, 2000).

8. W. Gilbert, "The RNA World," *Nature* 319 (1986): 618. Also, G. F. Joyce, "The Antiquity of RNA-Based Evolution," *Nature* 418 (2002): 214–221. But complexity theorist Stuart Kauffman has a different take in *Reinventing the Sacred: A New View of Science, Reason, and Religion* (New York: Perseus, 2008).

9. Stuart Kauffman contributed much of the groundwork in this field. See S. A. Kauffman, "Autocatalytic Sets of Proteins," *Journal of Theoretical Biology* 119 (1986): 1–24. Also, S. A. Kauffman, *The Origins of Order* (Oxford: Oxford University Press, 1993). For a recent review, see Wim Hordijk, Jotun Hein, and Mike Steel, "Autocatalytic Sets and the Origin of Life," *Entropy* 12 (2010): 1733–1742.

10. Degrees of homeostasis vary widely among cells and organisms. Many animals, especially vertebrates, maintain precise regulation of water and solutes in blood and tissue fluids. Most plants are imprecise regulators of sap contents and acid-base balance. See S. G. Pallardy, *Physiology of Woody Plants,* 3rd. ed. (London: Academic Press, 2008).

11. The concept was originally stated in Gilbert (1986). A *Scientific American* interview in 2011 with the eminent origin-of-life theorist Christian de Duve summarizes recent thinking on the emergence of the RNA world. See http://www.scientificamerican.com/podcast/episode

.cfm?id=science-legend-christian-de-duve-11–09–09. A short extension of these views is contained in C. de Duve, "Life as a Cosmic Imperative," *Philosophical Transactions of the Royal Society of London A* 369, no. 1936 (2011): 620–623. For more technical discussions, see H. Rauchfuss, trans. T. N. Mitchell, *Chemical Evolution and the Origin of Life* (Berlin: Springer-Verlag, 2008), 340.

12. Broad Institute (Harvard and MIT, 2016), CRISPR Timeline, https://www.broadinstitute.org/what-broad/areas-focus/project-spotlight/crispr-timeline.

13. Multiple articles and authors, 2015. "Customized Human Genes: New Promises and Perils," *Scientific American* (Nature America), Special Report, December, 1, 2015.

14. See for example, M. B. Simakov, "Asteroids and the Origin of Life—Two Steps of Chemical Evolution on the Surface of These Objects," *Earth Planets Space* 60 (2008): 75–82. Also, K. E. Smith, M. P. Callahan, P. A. Gerakines, J. P. Dworkin, and C. H. House, "Investigation of Pyrimidine Carboxylic Acids in CM2 Carbonaceous Chondrites: Potential Precursor Molecules for Ancient Coenzymes," *Geochimica et Cosmochimica Acta* 136 (2014): 1–12, http://science.gsfc.nasa.gov/691/cosmicice/reprints/Nictonic_acid_GCA_2014.pdf.

15. T. A. Lincoln and G. F. Joyce, "Self-Sustained Replication of an RNA Enzyme," *Science* 323, no. 5918 (2009): 1229–1232. Also, for more background discussion, see G. F. Joyce, "Evolution in an RNA World," *Cold Spring Harbor Symposia on Quantitative Biology* 74 (2009): 17–23.

16. David Deamer, *First Life: Discovering the Connections between Stars, Cells, and How Life Began* (Berkeley: University of California Press, 2011). This recent summary of Deamer's work on the evolution of primordial functional cells details his lipid work and extends thinking on life's debut on Earth to reasonable speculation in astrobiology. Also see for example, D. Deamer and A. Pohorille, "Synthetic Cells and Life's Origins," *Astrobiology* 8, no. 2 (2008): 453–455.

17. These conditions were partly simulated in the famous experiments by Stanley Miller, in which a variety of organic compounds vital to life, including amino acids, formed from some of the simplest small molecules—methane, ammonia, hydrogen, and so on—thought to be available in the Earth's earliest oceans. S. L. Miller, "Production of Amino Acids under Possible Primitive Earth Conditions," *Science* 117, no. 3046 (1953): 528–529. See also, J. Bada and A. Lazcano, "Prebiotic Soup—Revisiting the Miller Experiment," *Science* 300, no. 5620 (2003): 745–746. Modern geochemists find that the ancient atmosphere may not have been as uniformly and strongly reducing as Miller assumed, but local environments, as today, would have featured such conditions. Also, some of the insights of modern research in prebiotic chemical evolution were partly anticipated in the early twentieth century by the Russian chemist Alexander Oparin and the astute British biologist J. B. S. Haldane.

18. See, for example, http://astrobiology.nasa.gov/articles/o-oreos-nanosatellite-success-in-orbit/.

19. Behe (2006).

20. Ibid.

21. William Paley, *Natural Theology: Or Evidences of the Existence and Attributes of the Deity, Collected from the Appearances of Nature*, ed. M. D. Eddy and D. Knight (Oxford: Oxford University Press, reprinted 2008 [1802]), 384. See also, R. Dawkins, *The Blind Watchmaker: Why the Evidence of Evolution Reveals a Universe without Design* (New York: Norton, 1986) (more recent editions are available).

22. See, for example, Y. Sowa and R. M Berry, "Bacterial Flagellar Motor," *Quarterly Review of Biophysics* 41, no. 2 (2008): 103–132. Also, H. C. Berg, "The Rotary Motor of Bacterial Flagella," *Annual Review of Biochemistry* 72 (2003): 19–54.

23. See any college textbook in general biology or biochemistry.

24. Original studies that revealed the essentially crude and imprecise regulation of action of the flagellum were performed by Daniel Koshland and his associates in the 1970s, and later

confirmed by others in observations and visualizations of improved accuracy. See, for example, J. L. Spudich and D. E. Koshland Jr., "Non-Genetic Individuality: Chance in the Single Cell," *Nature* 262, no. 5568 (1976): 467–471. Also, S. C. Kuo and D. E. Koshland Jr., "Multiple Kinetic States for the Flagellar Motor Switch," *Journal of Bacteriology* 171, no. 11 (1989): 6279–6287. Also, C. V. Gabel and H. C. Berg, "The Speed of the Flagellar Rotary Motor of *Escherichia coli* Varies Linearly with Protonmotive Force," *PNAS* 100, no. 15 (2003): 8748–8751. Also, J. Xing, F. Bai, R. Berry, and G. Oster, "Torque–Speed Relationship of the Bacterial Flagellar Motor," *PNAS* 103, no. 5 (2006): 1260–1265.

Chapter 4: Ecology Emergent

Epigraph. Lauren Eiseley, *The Immense Journey* (NY: Vintage Books, Random House, 1957).

1. F. D. Ciccarelli, T. Doerks, C. von Mering, C. J. Creevey, B. Snel, and P. Bork, "Toward Automatic Reconstruction of a Highly Resolved Tree of Life," *Science* 311 no. 5765 (2006): 1283–1287. Also, C. Zimmer, http://blogs.discovermagazine.com/loom/2006/03/03/tree-of-life-c-2006/.

2. M. L. Sogin, H. G. Morrison, J. A. Huber, D. M. Welch, S. M. Huse, P. R. Neal, J. M. Arrieta, and G. J. Herndl, "Microbial Diversity in the Deep Sea and the Underexplored 'Rare Biosphere,'" *PNAS* 103, no. 32 (2006): 12115–12121.

3. D. B. Rusch, A. L. Halpern, G. Sutton, K. B. Heidelberg, S. Williamson, S. Yooseph, J. C. Venter et al., "The *Sorcerer II* Global Ocean Sampling Expedition: Northwest Atlantic through Eastern Tropical Pacific" (2007), http://www.plosbiology.org/article/info%3Adoi%2F10.1371%2Fjournal.pbio.0050077. Also, J. Craig Venter Institute, "Millions of New Genes, Thousands of New Protein Families Found in Ocean Sampling Expedition," *ScienceDaily*, March 14, 2007, www.sciencedaily.com/releases/2007/03/070314074922.htm.

4. See E. Singer, "The Ocean's Unforeseen Genomic Bounty," *MIT Technology Review* (2007), http://www.technologyreview.com/news/407517/the-oceans-unforeseen-genomic-bounty/.

5. Lynn Margulis, *Origin of Eukaryotic Cells* (New Haven, CT: Yale University Press, 1970); Lynn Margulis and Dorion Sagan, *Microcosmos: Four Billion Years of Evolution from Our Microbial Ancestors* (New York: HarperCollins, 1987); Lynn Margulis and Dorion Sagan, *Acquiring Genomes: A Theory of the Origins of Species* (New York: Perseus Books, 2002).

6. Recently, researchers have aligned the basic mitochondrial genome with a very close match in a type of small marine bacterium. See J. Cameron Thrash, Alex Boyd, Megan J. Huggett, Jana Grote, Paul Carini, Ryan J. Yoder, Barbara Robbertse, Joseph W. Spatafora, Michael S. Rappé, and Stephen J. Giovannoni, "Phylogenomic Evidence for a Common Ancestor of Mitochondria and the SAR11 Clade," *Scientific Reports*, PMC (2011) http://www.ncbi.nlm.nih.gov/pmc/articles/PMC3216501/.

7. A. Knoll, *Life on a Young Planet* (Princeton, NJ: Princeton University Press, 2003). Andrew Knoll's book (chapter 8) includes a lucid summary in accessible language of some of the most plausible cellular scenarios on the pathway of eukaryotic emergence. Also, W. Martin and M. Muller, "The Hydrogen Hypothesis for the First Eukaryote," *Nature* 392, no. 6671 (1998): 37–41.

8. See, for example, S. K. Hansen, P. B. Rainey, J. A. Haagensen, and S. Molin, "Evolution of Species Interactions in a Biofilm Community," *Nature* 445, no. 7127 (2007): 533–536.

9. See, for example, http://commonfund.nih.gov/hmp/. Medical researchers have begun a large-scale effort to understand the human microbiome, consisting of a huge variety of primarily mutualistic bacteria organized in complex communities at a number of key locations in our bodies, and increasingly confirmed to support human health. The Human Microbiome Project, coordinated by the US National Institutes of Health, has become a global leader in this endeavor.

10. S. H. D. Haddock, M. A. Moline, and J. F. Case, "Bioluminescence in the Sea," *Annual Review of Marine Science* 2 (2010): 443–493.

11. M. J. McFall-Ngai and M. K. Montgomery, "The Anatomy and Morphology of the Adult Bacterial Light Organ of *Euprymna scolopes* Berry (Cephalopoda: Sepiolidae)," *Biological Bulletin* 179 (1990): 332–339. Also, M. K. Montgomery and M. J. McFall-Ngai, "Embryonic Development of the Light Organ of the Sepiolid Squid *Euprymna scolopes* Berry," *Biological Bulletin* 184 (1993): 296–308. Also, K. J. Boettcher, E. G. Ruby, and M. J. McFall-Ngai, "Bioluminescence in the Symbiotic Squid *Euprymna scolopes* Is Controlled by a Daily Biological Rhythm," *Journal of Comparative Physiology A* 179 (1996): 65–73.

12. T. A. Koropatnick, J. T. Engle, M. A. Apicella, E. V. Stabb, W. E. Goldman, and M. J. McFall-Ngai, "Microbial Factor-Mediated Development in a Host-Bacterial Mutualism," *Science* 306, no. 5699 (2004): 1186–1188, doi:10.1126/science.1102218. Also M. McFall-Ngai, "The Importance of Microbes in Animal Development: Lessons from the Squid-Vibrio Symbiosis," *Annual Review of Microbiology* 680 (2014): 177–194.

13. O. K. Okamoto, L. Liu, D. L. Robertson, and J. W. Hastings, "Members of a Dinoflagellate Luciferase Gene Family Differ in Synonymous Substitution Rates," *Biochemistry* 40 (2001): 15862–15868. Also, D. H. Lee, M. Mittag, S. Sczekan, D. Morse, and J. W. Hastings, "Molecular Cloning and Genomic Organization of a Gene for Luciferin-Binding Protein from the Dinoflagellate *Gonyaulax polyedra*," *Journal of Biological Chemistry* 268 (1993): 8842–8850.

14. The relationship of zooxanthellae to limestone production in hermatypic corals has been known for several decades. See for example, S. V. Smith, and D. W. Kinsey, "Calcium Carbonate Production, Coral Reef Growth, and Sea Level Change," *Science* 194 (1976): 937–939.

15. The classic model of coral reef growth and development described in this paragraph was formulated largely out of the Caribbean reef studies of Thomas F. Goreau in the 1960s and 1970s and has been generally affirmed by most marine biologists since then. See T. F. Goreau, N. I. Goreau, and T. J. Goreau, "Corals and Coral Reefs," *Scientific American* 241, no. 2 (1979): 124–136.

16. Patterns of development of communities through ecological succession include increasing biodiversity, higher plant productivity and biomass storage in food chains, greater efficiency in recycling vital chemical elements, and so forth. These basic ecological phenomena are described in any college-level textbook in ecology.

17. E. O. Wilson, "The Species Equilibrium," in *Diversity and Stability in Ecological Systems*, Brookhaven Symposia in Biology No. 22 (Upton, NY: Brookhaven National Laboratory, 1969), 38–47.

18. See, for example, James Lovelock, *The Ages of Gaia: A Biography of Our Living Earth* (New York: Norton, 1995). Also, J. W. Kirchner, "The Gaia Hypothesis: Conjectures and Refutations," *Climatic Change* 58, no. 1 (2003): 21–45.

Chapter 5: Intimate Ark

Epigraph. Sarah Blaffer Hrdy, *Mother Nature: A History of Mothers, Infants, and Natural Selection* (NY: Pantheon, 1999).

1. A. J. F. Griffiths, W. M. Gelbart, J. H. Miller, and R.C. Lewontin, *Modern Genetic Analysis* (New York: W. H. Freeman, 1999), http://www.ncbi.nlm.nih.gov/books/NBK21351/.

2. These long-understood basic processes and principles in eukaryotic cell biology are discussed in detail in any college-level, general biology textbook.

3. A defective allele in one context may have survival value in another. The sickle-cell allele affects red blood cell (RBC) shape and oxygen-carrying capacity. It arose in populations living in a region with a high incidence of malaria. Nonmutant people with two "normal" alleles conferring the typical toroid shape of RBCs are severely debilitated or die from malaria; those with

both cell-shaping alleles coding for the mutant sickling condition are unable to carry normal levels of oxygen in their blood, with severe effects on their metabolism and endurance under physical stress. However, individuals with one normal and one sickle-trait allele (carriers of the mutation) are the best survivors. They resist malaria effectively and maintain a fairly robust level of physical fitness in general. Thus, the sickle-cell mutation persists indefinitely in populations as long as they are living with malaria as a "selection pressure."

4. H. T. Spieth, "Hawaiian Honeycreeper, *Vestiaria coccinea* (Forster), Feeding on Lobeliad Flowers, *Clermontia arborescens* (Mann) Hillebr.," *American Naturalist* 100 (1966): 470–473.

5. C. Darwin, *The Origin of Species* (1859, reprinted excerpts), in *Darwin: A Norton Critical Edition,* ed. P. Appleman (New York: Norton, 1970), 126.

6. E. C. Zimmerman, "Adaptive Radiation in Hawaii with Special Reference to Insects," *Biotropica* 2, no. 1 (1970): 32–38.

7. C. Packer, D. Scheel, and A. Pusey, "Why Lions Form Groups: Food Is Not Enough," *American Naturalist* 136, no. 1 (1990): 1–19. Also, A. Mosser and C. Packer, "Group Territoriality and the Benefits of Sociality in the African Lion, *Panthera leo,*" *Animal Behaviour* 78 (2009): 359–370. Also, K. L. VanderWaal, A. Mosser, and C. Packer, "Optimal Group Size, Dispersal Decisions and Postdispersal Relationships in Female African Lions," *Animal Behaviour* 77 (2009): 949–954.

8. A. Tennyson, *In Memoriam, A.H.H.,* Canto 56 (1850), http://www.theotherpages.org /poems/books/tennyson/tennyson04.html.

9. J. Roughgargen, *Evolution's Rainbow: Diversity, Gender, and Sexuality in Nature and People* (Berkeley: University of California Press, 2004).

10. See Plato, *Symposium,* trans. Alexander Nehamas and Paul Woodruff (Indianapolis: Hacket Publishing, 1989). For an exhaustive study of the topic of homosexuality in ancient Greece, see K. J. Dover, *Greek Homosexuality* (Cambridge, MA: Harvard University Press, 1989).

11. R. J. Morris, "Aikane: Accounts of Hawaiian Relationships in the Journals of Captain Cook's Third Voyage to Hawaii (1776–1780)," *Journal of Homosexuality* 19, no. 4 (1990): 21–54. See also, M. Coombs, "Commentary: Symposium Volume: Intersections: Sexuality, Cultural Traditions and the Law," *Yale Journal of Law and the Humanities* 241 (1996): 1–16.

Chapter 6: Social Order in Nature

Epigraph. Charles Darwin, *The Descent of Man, and Selection in Relation to Sex,* 2nd ed. (New York: D. Appleton and Company, 1898).

1. R. L. Caldwell and H. Dingle, "Stomatopods," *Scientific American* 234, no. 1 (1976): 80–89. This is still the best introduction to the behavioral ecology of mantis shrimps.

2. J. R. A. Taylor and S. N. Patek, "Ritualized Fighting and Biological Armor: The Impact Mechanics of the Mantis Shrimp's Telson," *Journal of Experimental Biology* 213 (2010): 3496–3504.

3. M. E. Harrington and G. Losey, "The Importance of Species Identification and Location on Interspecific Territorial Defense by the Damselfish, *Stegastes fasciolatus,*" *Environmental Biology of Fishes* 27, no. 2 (1990): 139–145.

4. A. Anker, G-V. Murina, C. Lira, J. A. V. Caripe, A. R. Palmer, and M-S. Jeng, "Macrofauna Associated with Echiuran Burrows: A Review with New Observations of the Innkeeper Worm, *Ochetostoma erythrogrammon* Leuckart and Rüppel, in Venezuela," *Zoological Studies* 44, no. 2 (2005): 157–190.

5. R. Bshary and R. Noë, "Biological Markets: The Ubiquitous Influence of Partner Choice on the Dynamics of Cleaner Fish—Client Reef Fish Interactions," in *Genetic and Cultural Evolution of Cooperation,* ed. P. Hammerstein (Cambridge, MA: MIT Press, 2003), chapter 9, 167–184.

6. C. Darwin, *The Origin of Species* (1859, reprinted excerpts), in *Darwin: A Norton Critical Edition,* ed. P. Appleman (New York: Norton, 1970), 236.

7. B. Holldobler and E. O. Wilson, *The Superorganism: The Beauty, Elegance, and Strangeness of Insect Societies* (New York: Norton, 2009), 544. This magisterial and accessible work is filled with wonders of emergence wrought by Darwinian evolution. Also, B. Holldobler and E. O Wilson, *The Ants* (Cambridge, MA: Harvard University Press, 1990).

8. P. W. Sherman, J. U. M. Jarvis, and S. H. Braude, "Naked Mole Rats," *Scientific American* 267, no. 2 (1992): 72–78.

9. See, for example, W. D. Hamilton, "The Evolution of Altruistic Behavior," *American Naturalist* 97, no. 896 (1963): 354–356. Also, J. Maynard-Smith, "Group Selection and Kin Selection," *Nature* 201 (1964): 1145–1147.

10. C. Darwin, *The Descent of Man, and Selection in Relation to Sex*, 2nd edition (New York: D. Appleton and Company, 1898), chapter 5, 132.

11. R. Dawkins, *The Selfish Gene* (New York: Oxford University Press, 1976), chapter 3.

12. The logical shortcomings of inclusive fitness are clearly described in M. A. Nowak and R. Highfield, *SuperCooperators: Altruism, Evolution, and Why We Need Each Other to Succeed* (New York: Free Press, 2011). And see the masterful recent synthesis of the emergence and consequences of eusociality in E. O. Wilson, *The Social Conquest of Earth* (New York: Norton, 2012); critical discussions of inclusive fitness appear through 142–146 and chapter 18.

13. M. A. Nowak, C. E. Tarnita, and E. O. Wilson, "The Evolution of Eusociality," *Nature* 466 (2010): 1057–1062. Also, Patrick Abbott et al., "Inclusive Fitness Theory and Evolution," *Nature* 471 (March 2011): E1–E4. This is a commentary signed by 137 academics defending inclusive fitness after the publication of the 2010 article by Nowak, Tarnita, and Wilson.

14. B. Allen, M. A. Nowak, and E. O. Wilson, "Limitations of Inclusive Fitness," *PNAS USA* 110, no. 50 (2013): 20135–20139. This article also appears in slightly modified form in E. O. Wilson, *The Meaning of Human Existence* (New York: Norton, 2014), appendix.

15. E. O. Wilson and B. Holldobler, "Eusociality: Origin and Consequences," *PNAS* 102 (2005): 13367–13371. Also, Nowak and Highfield (2011). Also, E. O. Wilson, "One Giant Leap: How Insects Achieved Altruism and Colonial Life," *BioScience* 58 (2008): 17–25.

16. J. E. Duffy and K. S. Macdonald, "Kin Structure, Ecology and the Evolution of Social Organization in Shrimp: A Comparative Analysis," *Proceedings of the Royal Society B: Biological Sciences* 277, no. 1681 (2010): 575–584. Also, J. E. Duffy, C. L. Morrison, and R. Rios, "Multiple Origins of Eusociality among Sponge-Dwelling Shrimps (*Synalpheus*)," *Evolution* 54, no. 2 (2000): 503–516.

17. D. S. Wilson and E. O. Wilson, "Evolution for the Good of the Group," *American Scientist* 96 (2008): 380–389 (386).

18. For example, see B. Heinrich, *Ravens in Winter* (New York: Summit Books, 1989). Also, K. Lorenz, *Here Am I—Where Are You? The Behavior of the Graylag Goose* (New York: Harcourt Brace Jovanovich, 1988).

19. R. Trivers, "The Evolution of Reciprocal Altruism," *Quarterly Review of Biology* 46 (1971): 35–57.

20. G. S. Wilkinson, "Reciprocal Food Sharing in the Vampire Bat," *Nature* 308 (1984): 181–184.

21. See, for example, R. Axelrod, *The Evolution of Cooperation*, rev. ed., foreword by Richard Dawkins (New York: Perseus, 2006).

22. See the discussion of cooperation manifested in distributed information processing (aka "group mind") at multiple levels in biology in D. S. Wilson and E. O. Wilson (2008).

23. Recent fossil finds in northwestern Africa (Chad) indicate that the evolutionary line leading to early humans split from that of the great apes close to seven million years ago. F. Guy, D. E. Lieberman, D. Pilbeam, M. Ponce de Leo, A. Likius, H. T. Mackaye, P. Vignaud, C. Zollikofer, and M. Brunet, "Morphological Affinities of the *Sahelanthropus tchadensis* (Late Miocene Hominid from Chad) Cranium," *PNAS* 102, no. 52 (2005): 18836–18841.

24. F. de Waal, *Primates and Philosophers: How Morality Evolved*, ed. and intro. by S. Machado and J. Ober (Princeton, NJ: Princeton University Press, 2006), 30–31.

25. C. McGraw, "Gorilla's Pet: Koko Mourns Kitten's Death," *Los Angeles Times*, January 10, 1985, http://articles.latimes.com/1985-01-10/news/mn-9038_1_pet-kitten.

26. de Waal (2006), 72. Also, F. de Waal, "On the Possibility of Animal Empathy," in *Feelings and Emotions: The Amsterdam Symposium*, ed. T. Manstead, N. Frijda, and A. Fischer (Cambridge: Cambridge University Press, 2003), 379–399. Recently evidence has begun to appear that crows and related birds may join various social mammals in possessing empathy and theory of mind. See, for example, N. J. Emery and N. S. Clayton, "Evolution of Intelligence in Corvids and Apes," *Science* 306 (2007): 1903–1907. Also, O. Fraser and T. Bugnyar "Do Ravens Show Consolation? Responses to Distressed Others," *PLoS One* 5, no. 5 (2010): e10605, http://www.klf.ac.at/downloads/Fraser%20&%20Bugnyar%202010.pdf.

27. E. Hatfield, R. L. Rapson, and L. Le Yen-Chi, "Emotional Contagion and Empathy," in *The Social Neuroscience of Empathy*, ed. J. Decety and W. Ickes (Cambridge, MA: MIT Press, 2009), 19–30. Also, de Waal (2006), 26.

28. N. Eisenberg, "Empathy and Sympathy," in *Handbook of Emotions*, ed. M. Lewis and J. M. Haviland-Jones, 2nd ed. (New York: Guilford Press, 2000), chapter 43, 677–691.

29. Roger T. Ames and Henry Rosemont Jr., *The Analects of Confucius: A Philosophical Translation* (New York: Ballantine Books, 1998), 189. *Shu* 恕 is often translated as reciprocity, but can also mean "empathy" as Huang Chun-chieh translates the term. See Huang Chun-chieh, *Humanism in East Asian Confucian Contexts* (Piscataway, NJ: Transcript-Verlag, 2010), 14. In modern Chinese, *shu* also means "to forgive."

30. *Mencius*, 7A4 (vol. 2), 265, quoted in Huang Chun-chieh (2010), 77.

31. Bryan W. Van Norden, *Mengzi: With Selections from Traditional Commentaries* (Indianapolis: Hackett, 2008), 46–47 (*Mencius*, 2A6, 3–4, or *Mengzi*, 2A6, 3–4).

32. J. Goodall, *Through a Window: My Thirty Years with the Chimpanzees of Gombe* (Boston: Houghton-Mifflin, 1990), 213. The contemporary Confucian scholar Huang Chun-chieh also points to examples of instinctive "morality" expressed in apes even crossing taxonomic boundaries. "This very consciousness has been witnessed in the animal kingdom. In recent years, female gorillas in zoos in Germany and the United States reportedly rescued human babies that had crawled into their cages, and protected them from other, less welcoming gorillas until humans could intervene." Huang Chun-chieh (2010), 75.

33. T. H. Huxley, "Evolution and Ethics" (The Romanes Lecture, 1893); the garden metaphor is developed in a follow-up address: "Evolution and Ethics: Prolegomena." Both published in *Collected Essays*, vol. 9 (London: Macmillan, 1894), available online, respectively, http://alepho.clarku.edu/huxley/CE9/E-E.html and http://alepho.clarku.edu/huxley/CE9/E-EProl.html. Also see J. G. Paradis and G. C. Williams, *Evolution and Ethics: T. H. Huxley's Evolution and Ethics with New Essays on its Victorian and Sociobiological Context* (Princeton, NJ: Princeton University Press, 1989).

34. T. H. Huxley, *Collected Essays*, vol. 9 (London: Macmillan, 1894), http://alepho.clarku.edu/huxley/CE9/E-EProl.html.

35. de Waal (2006), 7.

36. Dawkins (1976), 3. Dawkins' emphasis.

37. C. Korsgaard, Comments, Part 2, in de Waal (2006), 116.

38. C. Darwin, (1898 [1871]), 98, 134. Darwin's emphasis.

39. See, for example, T. Izumo, M. Shoraku, and N. Senryu (1748, putative authors), *Chushingura (The Treasury of Loyal Retainers)*, trans. Donald Keene (New York: Columbia University Press, 1971).

40. For an explanation of the Russian doll model, see de Waal in *Feelings and Emotions* (2003). Also de Waal (2006), 39, figure 4.

41. Mirror neurons are specialized nerve cells grouped in the motor region of primates' brains. The cells process information to translate both awareness of one's own actions and an affective internalization of actions, postures, facial expressions, and the like that we see in others. See, for example, B. Thomas, "What's So Special about Mirror Neurons?" *Scientific American Blogs*, November 6, 2012, http://blogs.scientificamerican.com/guest-blog/2012/11/06/whats-so-special-about-mirror-neurons/. Also, M. Iacoboni, *Mirroring People: The New Science of How We Connect with Others* (New York: Farrar, Straus, Giroux, 2008).

42. D. S. Wilson, *The Neighborhood Project: Using Evolution to Improve My City, One Block at a Time* (New York: Little, Brown, 2011).

43. E. O. Wilson, *Biophilia* (Cambridge, MA: Harvard University Press, 1984).

44. J. G. Speth, *Red Sky at Morning: America and the Crisis of the Global Environment* (New Haven, CT: Yale University Press, 2004).

Chapter 7: Self within World

Epigraphs. Antonio Porchia, *Voices* (*Voces*), trans. W. S. Merwin (Port Townsend, WA: Copper Canyon Press, 2003); and Friedrich Nietzsche, *The Gay Science* (Section 109), trans. Walter Kaufmann (New York: Random House, 1974), 169.

1. Philip J. Ivanhoe and Bryan W. Van Norden, *Readings in Classical Chinese Philosophy* (Indianapolis: Hackett, 2001), xiii.

2. Joseph Needham, *Science and Civilisation in China*, vol. 2 (Cambridge: Cambridge University Press, 1991), 338. As one of the oldest books in the world, the *Yijing* has represented an enigma to many Western philosophers, but one aspect that most can agree upon is that the human being is in essence a natural being and as such reflects the recurring patterns in nature. Historically, the most important philosopher and mathematician to consider the *Yijing* was Gottfried Wilhelm von Leibniz, who wrote its first European commentary in 1703. See Eric S. Nelson, "The Yijing and Philosophy: From Leibniz to Derrida," *Journal of Chinese Philosophy* 38, no. 3 (2011): 379. Also see Richard J. Smith, *Fathoming the Cosmos and Ordering the World: The Yijing (I Ching, or Classic of Changes) and Its Evolution in China* (Charlottesville: University of Virginia Press, 2008), 204.

3. Both these Daoist quotes are from the *Daodejing* and the *Zhuangzi*. See *Daodejing "Making This Life Significant": A Philosophical Translation*, trans. Roger T. Ames and David L. Hall (New York: Ballantine, 2003), 173; and *Chuang-tzŭ: The Seven Inner Chapters and Other Writings from the Book Chuang- tzŭ*, trans. A. C. Graham (London: George Allen & Unwin, 1981), 118.

4. Ku-ying Ch'en, trans. Roger T. Ames and Rhett Y. W. Young, *Lao Tzu: Text, Notes, and Comments* (San Francisco: Chinese Materials Center, 1977), 14.

5. This is the translation of *dao* used by Roger T. Ames and David L. Hall in their *Daodejing "Making This Life Significant"* (2003).

6. Ibid., 214.

7. Ibid., 134.

8. See Culliney and Jones, "Confucian Order at the Edge of Chaos," *Zygon: Journal of Religion and Science* 33, no. 3 (1998): 395–404.

9. C. G. Langton, "Computation at the Edge of Chaos," *Physica D* 42 (1990): 35.

10. Ames and Hall (2003), 142.

11. A. C. Graham, *Disputers of the Tao: Philosophical Arguments in Ancient China* (Chicago: Open Court, 1989), 325.

12. Graham (1981), 151.

13. J. Gleick, *Chaos: Making a New Science* (New York: Penguin Books, 1987), 233.

14. S. Kauffman, *Reinventing the Sacred* (New York: Basic Books, 2008), 23.

15. Ames and Hall (2003), 139.

16. Ibid., 143.

17. See Ch'en (1977), 208, n. 5, for further discussion.

18. Ames and Hall (2003), 142 (chapter 42).

19. In J. Baird Callicott and Roger T. Ames, eds., *Nature in Asian Traditions of Thought: Essays in Environmental Philosophy* (Albany: State University of New York Press, 1989), 70.

20. Callicott and Ames (1989), 70.

21. Graham (1981), 53.

22. Francis H. Cook, *Hua-yen Buddhism: The Jewel Net of Indra* (University Park: Pennsylvania State University Press, 1977), 2.

23. A hologram is, according to Merriam Webster, "a three-dimensional image reproduced from a pattern of interference produced by a split coherent beam of radiation (as a laser); *also:* the pattern of interference itself."

24. Thich Nhat Hanh, *Zen Keys: A Guide to Zen Practice* (Garden City, NJ: Doubleday, 1974), 87.

25. For convenience sake we will use Sanskrit spellings in English for Buddhist terms although most appear originally in Pali.

26. John M. Koller, *Asian Philosophies,* 4th ed. (Upper Saddle River, NJ: Prentice Hall, 2002), 167.

27. Hanh (1974), 124.

28. Ibid., 129. Italics in the original. Zen masters employ a number of strategies (such as the koan) to bring about "crises" that challenge the comfort we enjoy being a self that has an abiding sense of the reality of a world "out there" accompanied by a sense of the me being somewhere "in here." As a Zen master, Thich Nat Hanh realizes that the idea of a distinct, separate, and substantial independent self is the greatest obstacle in overcoming suffering and is the cause of alienation from all that is.

29. Francis Crick, *The Astonishing Hypothesis: The Scientific Search for the Soul* (New York: Simon and Schuster, 1994), 33.

30. Our numbering follows the Diels-Kranz's numbering of the fragments. Marcovich has this fragment numbered 40c. The Greek spelling of Herakleitos is used throughout instead of the typical version of Heraclitus. All Greek translations are David Jones' unless otherwise specified. For a standard collection of Presocratic philosophy see G. S. Kirk and J. E. Raven, *The Presocratic Philosophers: A Critical History with a Selection of Texts* (London: Cambridge University Press, 1957).

31. Fragment 10. Diels-Kranz.

32. Fragment 67. Diels-Kranz.

33. Fragment 50. Diels-Kranz.

34. Fragment 101. Diels-Kranz.

35. Fragment 45. Diels-Kranz.

36. The quote is "The safest general characterization of the European philosophical tradition is that it consists of a series of footnotes to Plato." Alfred North Whitehead, *Process and Reality* (New York: Free Press, 1979), 39.

37. See Paul Guyer, *Critique of the Power of Judgment* (Cambridge: Cambridge University Press, 2000), 234; or 5:360 in Kant's "Critique of the Teleological Power of Judgment."

Chapter 8: From Self to Sage

Epigraph. Quoted from A. C. Graham, *Disputers of the Tao: Philosophical Arguments in Ancient China* (Chicago: Open Court, 1989), 101. The *Guanzi's Nèiyè* 內業, "Inner Training or Inner Life," chapter is one of the earliest texts depicting Daoist self-cultivation training and techniques, including meditation practices. The widely accepted theory has the "proto-*Guanzi*," the core of the text we have today, being compiled around 250 BCE. See W. Allyn Ricket, *Guanzi: Political, Economic, and Philosophical Essays from Early China* (Boston: Cheng and Tsui Company, 2001), 15.

1. Chun-chieh Huang, *Humanism in East Asian Confucian Contexts* (Piscataway, NJ: Transcript-Verlag, 2010), 77. Huang follows convention by translating *ren* as "benevolence," which is certainly not wrong, but may be somewhat misleading for the English reader. Roger Ames, in *Confucian Role Ethics: A Vocabulary* (Honolulu: University of Hawai'i Press, 2011, 165), translates the term in a variety of ways, sometimes as "relational virtuosity," which we like considerably, and "consummate person or conduct" depending on context (179). Huang's language here of "substance" and "principle" may lend itself to an understanding of the nature of *ren* as being opposite of what we're arguing for in this chapter.

2. This etymology is reinforced by the alternative graph (a pregnant body, *shen* 身, over a heart-mind, *xin* 心, found in the Guodian texts. What this suggests is a woman with child is the ultimate expression of intimate relationality. We are grateful to Roger T. Ames for pointing this out to us, and for his other suggestions throughout this chapter.

3. Ames (2011), 96.

4. Huang (2010), 78. Huang again follows convention in his translation of *cheng* 誠 as sincerity. Etymologically, the character *cheng* consists of "completing" or "becoming" (*cheng* 成) and "saying" or "discoursing" (*yan* 言). What this suggests is that the sage's accomplishment resides in the ability to communicate effectively and collaboratively with others, which results in a shared co-creativity in any given situation. Here we follow Roger Ames' analysis of the character. See Ames (2011), 127.

5. Ames (2011), 138. (*Mencius*, Book 2A6.5 or *Mengzi*, Book 2A6.5). The translation has been emended slightly. Ames opts to translate *siduan* 四端 as the "four inklings" instead of "sprouts," to likely retain a more humanistic sense of the term. *Si* 四 means the number "four" and *duan* 端 is the word emended here as sprouts. We have also emended "pity" from the quote with "compassion."

6. Huang (2010), 78.

7. Ames (2011), 177.

8. Ibid., 85.

9. Roger T. Ames and David L. Hall, *Daodejing "Making This Life Significant": A Philosophical Translation* (New York: Ballantine Books, 2003), 122.

10. Ibid., 151.

11. Ibid., 214.

12. A. C. Graham states that "The word *shen* tends to be used as a stative verb rather than a noun, of mysterious power and intelligence radiating from a person or thing; as English equivalent we choose 'daimonic.' Man himself can aspire . . . to that supremely lucid awareness which excites shudder of numinous awe. Thus the sage in *Chuang Tzu [Zhuangzi]* is the 'daimonic man.'" We follow Graham's lead because the etymological connection in English between "daimonic" and its Greek root *daimōn*, which contains the multiple meanings of god, intermediate spirit, one's character, and one's genius. In its adjectival form, the term means being skilled in a thing. For the Daoist, being a sage means to be skilled at becoming the world. See A. C. Graham, *Disputers of the Tao: Philosophical Arguments in Ancient China* (Chicago: Open Court, 1989), 101.

13. Ku-ying Ch'en, *Lao Tzu: Text, Notes, and Comments*, trans. Roger T. Ames and Rhett Y. W. Young (San Francisco: Chinese Materials Center, 1977), 14.

14. Brook Ziporyn, trans., *Zhuangzi: The Essential Writings* (Indianapolis: Hackett, 2009), 22.

15. Ibid., 22.

16. The Maori word *whānau* can mean family in the broadest sense: blood relatives, those adopted, and those intentionally chosen to be part of a particular intimate circle of people.

17. Steven Squyres, *Roving Mars: Spirit, Opportunity, and the Exploration of the Red Planet* (New York: Hyperion, 2005).

18. Graham (1981), 281.

19. *Pu* (樸), unworked, unhewn wood, or uncarved block, is used as an image of Dao in the *Dao De Jing* and represents the spontaneous unfolding of the universe or *ziran*; it can be seen as a metaphor for *wuwei*. Following Zhuangzi's idea that *"Dao comes about as we walk it,"* our suggestion is that the sage plays a responsive, active, and ultimately vital role in the universe's potential transforming.

20. Graham (1989), 191.

21. Ames and Hall (2003), 175. Italics added.

22. Graham (1981), 16.

23. Ibid., 156.

24. Thomas P. Kasulis, *Intimacy or Integrity: Philosophy and Cultural Difference* (Honolulu: University of Hawai'i Press, 2002), 102.

25. For Nietzsche on "how the 'true' world became a fable," see Nietzsche, *Twilight of the Idols: or How to Philosophize with a Hammer*, trans. Duncan Large (New York: Oxford University Press, 2009). For Heidegger on "the world-hood of the world" see Heidegger, *Being and Time: A Revised Edition of the Stambaugh Translation*, trans. Joan Stambaugh and rev. Dennis J. Schmidt (Albany: State University of New York Press, 2010). And for Merleau-Ponty on the "flesh of the world" see Merleau-Ponty, *The Visible and Invisible*, trans. Alphonso Lingis (Chicago: Northwestern University Press, 1969).

26. Richard Rhodes, *Dark Sun: The Making of the Hydrogen Bomb* (New York: Simon and Schuster, 1995). This volume is the centerpiece of Rhodes' trilogy on the history of nuclear weaponry. He weaves the complex and compellingly written tale of the H-bomb—research, testing, politics, and espionage—from the American-British perspective and with extensive Soviet archives that were newly opened in the 1990s under glasnost.

27. Kasulis (2002), 61.

28. Ibid.

29. Ibid., 62. Although Kasulis more faithfully represents earlier features of existentialism, his analysis does tend to overlook the more intimacy based orientations of later thinkers such as Martin Heidegger and Maurice Merleau-Ponty.

30. Ibid., 63.

31. Kazuaki Tanahashi, *Moon in a Dewdrop: Writings of Zen Master Dōgen* (New York: North Point Press, 1985), 163.

Chapter 9: From Self to No-Self to All-Self

Epigraphs. Johann Christian Friedrich Hölderlin, *Werke und Briefen*, Band I, 300; and Kazuaki Tanahashi, *Moon in a Dewdrop: Writings of Zen Master Dōgen* (New York: North Point Press, 1985), 164.

1. Kasulis adds that knowledge "is embodied in persons embedded within their intimate community of praxis and knowledge. Studying the world entails studying oneself as well; it begins not with a polarity but in an *in medias res*. . . . In the intimacy model, . . . knowledge occurs where world and self intersect." Thomas P. Kasulis, *Intimacy or Integrity: Philosophy and Cultural Difference* (Honolulu: University of Hawai'i Press, 2002), 103.

2. See for example, W. S. Hylton, "Craig Venter's Bugs Might Save the World," *New York Times,* May 30, 2012, http://www.nytimes.com/2012/06/03/magazine/craig-venters-bugs -might-save-the-world.html?pagewanted=all&_r=0. Also, N. Annaluru, H. Muller, L. Mitchell, S. Ramalingam, G. Stracquadanio, S. Richardson, J. Boeke, and seventy-four others, "Total Synthesis of a Functional Designer Eukaryotic Chromosome," *Science* 344, no. 6179 (2014): 55–58.

3. Understanding of this sense of domination traces back to 1967. See Lynn White's landmark article, "The Historical Roots of Our Ecologic Crisis," in *Science* 155, no. 3767 (March 10, 1967): 1203–1212. White's position is well represented in *The Stanford Encyclopedia of Philosophy* entry by Andrew Brennan and Yeuk-Sze Lo, "Environmental Ethics," *The Stanford Encyclopedia of Philosophy* (Fall 2011 ed.), ed. Edward N. Zalta, http://plato.stanford.edu /archives/fall2011/entries/ethics-environmental/. "According to White, the Judeo-Christian idea that humans are created in the image of the transcendent supernatural God, who is radically separate from nature, also by extension radically separates humans themselves from nature. This ideology further opened the way for untrammelled exploitation of nature." "Modern Western science itself," White argues, "was 'cast in the matrix of Christian theology' so that it too inherited the 'orthodox Christian arrogance toward nature'" (1967, 1207).

4. Kasulis (2002), 77.

5. David V. Erdman, ed., *The Complete Poetry & Prose of William Blake* (New York: Anchor Books, 1988), 490.

6. Indra's Net is developed most fully in the Mahāyāna sutra (or thread) known as the *Avatamsaka Sutra,* or *Flower Garland Sutra,* and provides us with a stunning depiction of reality. Why Indra, a warrior god found in the *Rig Veda,* is used in this sutra perhaps represents our ongoing struggle against the seductive forces that lead us away from realizing the fundamental relationality of who and what we are and what we must become in order to affect a greater sensitivity to what we're ultimately a part of. When Buddhism made its way to China, Indra's Net formed a perfect metaphor for their conception of *wanwu* 萬物, which literally means "the ten thousand things," a large number used in ancient times for all that is happening within the totality of what is. *Wanwu* is similar to the Greek *murios* (μυρίοι), which also means the number ten thousand, or endless. Μυρίοι is where we get our word "myriad" from; hence it often drives the translation of *wanwu* as the "myriad things."

7. Roger T. Ames and Henry Rosemont Jr., *The Analects of Confucius: A Philosophical Translation* (New York: Ballantine, 1998), 110 (see 6.30). We have slightly modified the translation by rendering *ren* as "consummate person" instead of "authoritative person" to get at the interconnectedness we are discussing more fully. This modification is not a superfluous one since Ames and Rosemont are now often using the term consummate as well. For more discussion on this understanding of Confucian thought, see Culliney and Jones, "Confucian Order at the Edge of Chaos," *Zygon: Journal of Religion and Science* 33, no. 3 (1998): 395–404.

8. Kasulis (2002), 103.

9. For an accessible rendition of Buddhist practice, see "The Foundations of Buddhism" chapter in the Dalai Lama's, *Essence of the Heart Sutra,* trans. Gesge Thupten Jinpa (Boston: Wisdom Publications, 2002). The Eightfold Path is discussed and contextualized on page 28 of the chapter.

10. Quoted from *The Dalai Lama: Essential Writings,* ed. Thomas A. Forsthoefel (Maryknoll, NY: Orbis Books, 2008), 69.

11. See David Hume's "On Personal Identity," in Hume, *Selections from An Enquiry Concerning Human Understanding and A Treatise of Human Nature,* Section 6 (La Salle, IL: Open

Court, 1966), 257. Hume also believed that morality was based on feelings rather than abstract principles. The German philosopher Immanuel Kant was prompted to remark that David Hume had awoken him from his "dogmatic slumber." Through Kant's "awakening," we have become in so many ways inheritors of his moral theory that moves us away from such feelings as empathy to the universalization of ethical principles. In Kant's case, this is the categorical imperative, which states one should "act only according to that maxim by which you can at the same time will that it be a universal law"(AK 4:402). AK is the standard abbreviation for the Berlin *Akademie* edition of Kant's complete works in the form of volume: page number. For another reference see Immanuel Kant, *Fundamental Principles of the Metaphysic of Morals*, trans. Thomas K. Abbott (Indianapolis: Bobbs-Merrill Educational Publishing, 1949).

12. Thich Nhat Hanh, *The Heart of Understanding: Commentaries on the Prajñaparamita Heart Sutra* (Berkeley, CA: Parallax Press, 1988), 7.

13. Hanh (1988), 7.

14. Makoto Ueda, *Matsuo Basho* (Tokyo: Kodansha International, 1982 [Twayne 1970]), 25.

15. Ibid.

16. Jane Reichhold, *Basho: The Complete Haiku* (Tokyo: Kodansha International, 2008), 77.

17. Ibid., 270.

18. Richard Rhodes, *Dark Sun: The Making of the Hydrogen Bomb* (New York: Simon and Schuster, 1995), 509.

19. Forsthoefel (2008), 156.

20. Ronald Pine, *Science and the Human Prospect* (Belmont, CA: Wadsworth, 1989), 243.

21. Robert Aitken, *A Zen Wave: Bashō's Haiku and Zen* (Washington DC: Shoemaker & Howard, 1978), 4.

22. Ibid., 5

23. Ibid., 4.

24. Ibid., 7.

25. Ibid., 6.

26. Hanh (1988), 3.

27. Ibid., 4.

28. Stuart Kauffman, *Reinventing the Sacred* (New York: Basic Books, 2008), 232.

29. Thich Nhat Hanh frequently uses the analogy of the cloud and rain in relation to the fear of death, and we borrow this analogy here.

30. Dōgen, *Shōbōgenzō*, cited in Kim Hee-jin, *Dōgen Kigen: Mystical Realist* (Tucson: University of Arizona Press, 1975), 97. For a collection of essays on Buddha-nature in this vein, see David Jones, ed., *Buddha Nature and Animality* (Fremont, CA: Jain Publishing, 2007).

31. Kazuaki Tanahashi, *Moon in a Dewdrop: Writings of Zen Master Dōgen* (New York: North Point Press, 1985), 164.

32. Thich Nhat Hanh, *The Diamond That Cuts through Illusion* (Berkeley, CA: Parallax Press, 1992), 35–36.

33. Nagarjuna, *Mulamadhyamakakarikas* (M 25: 19–20), translated in Frederick I. Streng, *Emptiness: A Study in Religious Meaning* (Nashville, TN: Abingdon Press, 1967).

34. See, for example, http://www.kumukahi.org/units/ke_ao_akua/akua/pele.

35. Tanahashi (1985), 70.

36. Hanh (1992), 25.

37. For a sense of the situation in Myanmar, see http://www.bbc.com/news/magazine-22356306; and for Bhutan and Sri Lanka, see http://www.thenation.com/article/174104/buddhist-violence-burma.

Chapter 10: Anti-sage

Epigraphs. Henry Kissinger, Kissinger Cables, via WikiLeaks, March 10, 1975. Donald J. Trump accepting his nomination for president by the Republican national convention in Cleveland, Ohio, July 21, 2016.

1. Occasionally it is possible to identify fragments of a complex, anti-sagely-inspired system that reemerge in a positive light in later institutional worldlines. One such example is German rocketry science, developed in World War II, that later contributed to early explorations of space by the United States and the Soviet Union. However, the same technology enabled the development of intercontinental ballistic weaponry.

2. An exceptionally clear account of the trajectory of an anti-sage appears in the critically acclaimed history of the People's Temple cult: T. Reiterman and J. Jacobs, *Raven: The Untold Story of the Rev. Jim Jones and His People* (New York: E. P. Dutton, 1982), 622; also, J. Sheeres, *A Thousand Lives: The Untold Story of Hope, Deception, and Survival at Jonestown* (New York: Free Press, 2011), which references volumes of documents long held confidential by the FBI.

3. R. J. Lifton, *Destroying the World to Save It* (New York: Henry Holt, 1999); see Lifton's seven characteristics of "world-destroying cults," 202–213.

4. J. Ginges, I. G. Hansen, and A. Norenzayan, "Religion and Popular Support for Suicide Attacks," *Psychological Science* 20 (2009): 224–230.

5. R. J. Lifton, *The Protean Self: Human Resilience in an Age of Fragmentation* (New York: Basic Books, 1993).

6. R. J. Lifton, quoted from an interview with Bill Moyers, *Moyers in Conversation*, PBS Television, September 17, 2001.

7. On March 9, 1954, Murrow, on his live CBS television program *See It Now*, took on McCarthy's witch hunts for their lack of evidence. This broadcast, in which Murrow emphasized that "we must not confuse dissent with disloyalty. . . . We will not be driven by fear into an age of unreason," has been seen as a turning point leading to McCarthy's downfall. See, for example, James McEnteer, *Shooting the Truth: The Rise of American Political Documentaries* (Westport, CT: Praeger, 2005), 7.

8. For example, Henry Kissinger—see Gary Bass, *The Blood Telegram: Nixon, Kissinger, and a Forgotten Genocide* (New York: Knopf, 2013); also, Oliver North—see Alexander Cockburn and Jeffrey St. Clair, *Whiteout: The CIA, Drugs, and the Press* (Brooklyn, NY: Verso, 1999); also, Richard Cheney—see Peter Baker, *Days of Fire: Bush and Cheney in the White House* (New York: Doubleday, 2013). Note, Baker's study suggests that President Bush resisted the manipulations of Mr. Cheney to a greater degree than indicated in other sources.

9. C. Johnson, *Nemesis: The Last Days of the American Republic* (New York: Henry Holt, 2006), 26–27.

10. Ibid., 29.

11. Mark Bowen, *Censoring Science: Inside the Political Attack on Dr. James Hansen* (New York: Dutton/Penguin, 2007).

12. See, for example, B. J. Bush, "Addressing the Regulatory Collapse behind the Deepwater Horizon Oil Spill: Implementing a 'Best Available Technology' Regulatory Regime for Deepwater Oil Exploration Safety and Cleanup Technology," *Journal of Environmental Law and Litigation* 26, no. 2 (2011): 535–569. Also, "Safety First, Fracking Second," *Scientific American*, November 2011. Also Ken Ward Jr., "Shafted: How the Bush Administration Reversed Decades of Progress on Mine Safety," *Washington Monthly*, March 2007, http://www.washingtonmonthly.com/features/2007/0703.ward.html.

13. For a thorough discussion on the relationship between money and the radical right, see Jane Mayer, *Dark Money: The Hidden History of the Billionaires behind the Rise of the Radical Right* (New York: Doubleday, 2016).

14. See *The Economist*, December 7, 2013, 30.

15. See R. Azlan, *Zealot: The Life and Times of Jesus of Nazareth* (New York: Random House, 2013), for a recent and well-researched alternative account of the historical Jesus.

16. See Galatians 2:6. According to Azlan (2013, 185), "Paul holds particular contempt for the Jerusalem-based triumvirate of James, Peter, and John, whom he derides as the 'so-called pillars of the church' (Galatians 2.9). 'Whatever they are makes no difference to me,' he writes. 'Those leaders contributed nothing to me' (Galatians 2.6). The apostles may have walked and talked with the living Jesus (or, as Paul dismissively calls him, 'Jesus-in-the-flesh')," but "Paul walks and talks with the divine Jesus: they have, according to Paul, conversations in which Jesus imparts secret instructions intended solely for his ears" (185). Unlike his counterparts in Jerusalem, Paul was an "educated Greek-speaking Diaspora Jew and [a] citizen in one of the wealthiest port cities [Tarsus] in the Roman Empire" (183) who "does repeatedly insist that he has witnessed the risen Jesus for himself" (184). Paul's focus was on the conversion of gentiles. He is largely responsible for initiating the claim that Jesus was the Christ (188). Paul's teachings and writings gained increasing influence in the evolution of Christianity from its Jewish roots to the official religion of Rome.

17. See Azlan's "Epilogue" (2013, 213–216) for a clear and brief discussion of the Council of Nicaea.

18. In the Hebrew Bible, see the Book of Joshua—accounts of campaigns against the Canaanites. Also the Book of Judges highlights Jewish aggression against neighboring tribes.

19. As Livia Kohn notes, "temples and practitioners are administered by the Bureau of Religious Affairs, which is staffed by hard-core Communists with little patience for religious needs or activities. With an increase in local travelers, foreign visitors, and Overseas Chinese—who were essential in the religious revival after 1980—Daoist institutions also came to be supervised by the Department of Tourism. Manned by foreign-oriented and more modern officials, the Department is interested mainly in revenue and the smooth entertainment of large crowds. It, too, has no concern with religious activities or the creation of a spiritual atmosphere" (186). Kohn, *Introducing Daoism* (New York: Routledge, 2009).

20. "An Unintegrated Life Is Not Worth Living," in *Confucius Now: Contemporary Encounters with the* Analects, ed. David Jones (Chicago: Open Court, 2008), 154.

21. Lifton (1993).

22. Herrlee Glessner Creel, *Confucius and the Chinese Way* (New York: Harper, 1949), 4.

23. Benjamin Ginsberg, *The Fall of the Faculty: The Rise of the All-Administrative University and Why It Matters* (New York: Oxford University Press, 2011), 24.

24. See ABC News online, http://abcnews.go.com/Business/penn-states-spanier-leads-top-11-highest-paid/story?id=19151598.

25. Ginsberg (2011), 24.

26. Jack Stripling, "Kerrey Says His Emeritus Role at New School Invites Controversy." *Chronicle of Higher Education* 59, no. 20 (January 15, 2013).

27. Ginsberg (2011), 92.

28. See, for example, R. Aviv, "The Imperial Presidency," *New Yorker*, September 9, 2013, 60–71.

29. NYU's Global Network University has campuses or "academic sites" in Accra, Abu-Dhabi, Berlin, Buenos Aires, Florence, London, Madrid, Paris, Prague, Shanghai, Singapore, Sydney, Tel-Aviv, and Washington, DC. See http://www.nyu.edu/global.html.

30. Aviv (2013), 69.

31. Ibid.

32. Ginsberg (2011), 25. Educational management now includes an army of "information technology specialists, counselors, auditors, accountants, admission officers, development officers, alumni relations officials, human resources staffers, editors and writers for school publications, attorneys, and a slew of others." See also pages 24 and 29.

33. Ernest L. Boyer, *Scholarship Reconsidered: Priorities of the Professoriate* (New York: John Wiley, 1990).

34. The potential emergence sometimes appears immediately or sometimes years later with a visit, an email, or letter from a former student who has realized some discovery or self-discovery at work or play, while traveling or at home, in a career as scientist, humanist or engineer, artist, teacher, or a nurse—who remembers the spark that ignited the curiosity or the fervor. There is no award or reward that feels better than this to a college instructor.

Chapter 11: Into Indra's Net

Epigraph. Arthur C. Clarke, Farewell address on his ninetieth birthday in Colombo, Sri Lanka, December 16, 2007, http://www.spaceref.com/news/viewsr.html?pid=26327.

1. The Freeman Foundation was established in 1994 by Houghton "Buck" Freeman and has been steadfastly dedicated to improving understanding of Asia by providing generous grants mostly for educators. The foundation remains family-run and is committed to the teaching of Asia studies in university classrooms. It has had its own fractal impact on education.

2. An interview by Alice Li Hagen, titled "International Education Leader—Dr. Joseph L. Overton," aired on *ThinkTech: Hawaii's Global Future* on October 29, 2015. The interview may be found at https://www.youtube.com/watch?v=wsVUE-XagXI.

3. Quoted from Roger T. Ames, *Confucian Role Ethics* (Hong Kong: The Chinese University Press, 2011), 126, 196.

4. *Strike Magazine*, August 17, 2013.

5. Carl Sagan, *The Demon-Haunted World: Science as a Candle in the Dark* (New York: Random House, 1996).

6. See http://www.huffingtonpost.com/2014/07/05/pope-francis-nature-environment-sin-_n_5559631.html. Also, http://www.telegraph.co.uk/earth/environment/climatechange/11107434/Scientists-turn-to-Pope-Francis-and-worlds-religions-to-save-the-planet.html.

7. A. Schiffrin and E. Kircher-Allen, eds., foreword by J. D. Sachs and introduction by J. E. Stiglitz, *From Cairo to Wall Street: Voices from the Global Spring* (New York: New Press, 2012).

8. In the early 1990s, biologists such as Paul R. Ehrlich and Edward O. Wilson began alerting the scientific community to the gravity of an ongoing collapse of global biodiversity. In 1993, Wilson published his estimate of the worldwide loss of some thirty thousand species per year that could be linked to human expansion and exploitation of the earth. E. O. Wilson, *The Diversity of Life* (Cambridge, MA: Harvard University Press, 1993). Until our species began its accelerating rise to global dominance, there had been five mass extinctions of life; the fifth extinguished the dinosaurs and much of the rest of planetary biodiversity some sixty-five million years ago. Now the sixth mass extinction, human-caused, confirmed by calculations and analysis of numerous other scientists, is rapidly progressing out of control. For a recent survey, see E. Kolbert, *The Sixth Extinction: An Unnatural History* (New York: Henry Holt and Company, 2014).

9. E. O. Wilson, *Biophilia* (Cambridge, MA: Harvard University Press, 1984).

10. The rivet model of ecological structure and function was proposed by Paul and Anne Ehrlich. Not all "rivets" may have equal holding power in an airframe, but as they are removed some will prove to be vital. See P. R. Ehrlich and A. H. Ehrlich, *Extinction: The Causes and Consequences of the Disappearance of Species* (New York: Random House, 1981).

11. For ESA, see http://www.fws.gov/endangered/laws-policies/; for CBD, see http://www.cbd.int/convention/.

12. B. J. Cardinale et al., "Biodiversity Loss and Its Impact on Humanity," *Nature* 486 (2012): 59–61.

13. See, for example, F. Danielsen, M. K. Sorensen, M. F. Olweg et al., "The Asian Tsunami: A Protective Role for Coastal Vegetation," *Science* 310 (2005): 643, http://www.sciencemag.org /content/310/5748/643.full.

14. G. M. Woodwell, "Resources and Compromise (Letter)," *Science* 229 (1985): 600.

15. S. M. Hsiang, K. C. Meng, and M. A. Cane, "Civil Conflicts Are Associated with the Global Climate," *Nature* 476 (2011): 438–441, http://www.nature.com/nature/journal/v476 /n7361/full/nature10311.html.

16. Peter Matthiessen, *In Paradise* (New York: Riverhead Head Books, 2014), 201; "'In Paradise,' Matthiessen Considers Our Capacity for Cruelty," interview by Tom Vitale, April 5, 2014, http://www.npr.org/books/authors/137941494/peter-matthiessen.

17. Peter Matthiessen, *The Snow Leopard* (New York: Bantam, 1979).

18. Ibid., 14–15.

19. Ibid., 20–21.

20. S. Pinker, *The Better Angels of Our Nature: Why Violence Has Declined* (New York: Viking, 2011).

21. H. Kahn, *On Escalation: Metaphors and Scenarios* (New York: Praeger, 1965; Piscataway, NJ: Transaction Publishers, 2009), 194.

22. Multiple articles and authors, 2015. "Customized Human Genes: New Promises and Perils," Special Report, *Scientific American* (Nature America), December, 1, 2015, http://www .scientificamerican.com/report/customized-human-genes-new-promises-and-perils/. Well-written general treatments of CRISPR technology and applications have also appeared in 2015 in *The Economist:* http://www.economist.com/news/briefing/21661799-it-now-easy-edit -genomes-plants-animals-and-humans-age-red-pen. Also see Michael Specter, "The Gene Hackers," *New Yorker,* November 16, 2015, 52–61.

23. Nick Bostrom, *Superintelligence: Paths, Dangers, Strategies* (London: Oxford University Press, 2014).

24. I. J. Good, "Speculations concerning the First Ultraintelligent Machine," *Advances in Computers* 6 (1965): 31–88.

25. J. Barbour, *The End of Time: The Next Revolution in Physics* (New York: Oxford University Press, 1999).

26. L. Smolin, *Time Reborn: From the Crisis in Physics to the Future of the Universe* (New York: Houghton-Mifflin Harcourt, 2013).

About the Authors

JOHN L. CULLINEY is a biologist educated at Yale and Duke Universities. His previous books have focused on topics ranging from ecology and human impacts on the continental shelf of North America to the environmental history of the Hawaiian Islands. He is currently professor emeritus of biology at Hawai'i Pacific University. He now makes his home in Volcano, Hawai'i.

DAVID JONES is professor of philosophy at Kennesaw State University in Atlanta, editor of the journal *Comparative and Continental Philosophy* (Taylor and Francis), and the series editor for the *Series on Comparative and Continental Philosophy* (Northwestern University Press). He has been visiting professor of Confucian classics at Emory and visiting scholar at the Institute for Advanced Studies in Humanities and Social Science at National Taiwan University. Most of his research and writing are in the areas of Chinese and Greek philosophy, but he has also published on Japanese, Buddhist, environmental, and continental philosophy.

Printed in the United States
By Bookmasters